Research Methodology in Marketing

Martin Eisend • Alfred Kuss

Research Methodology in Marketing

Theory Development, Empirical Approaches and Philosophy of Science Considerations

Springer

Martin Eisend
Faculty of Business Administration and Economics
European University Viadrina
Frankfurt (Oder), Germany

Alfred Kuss
Marketing Department
Freie Universität Berlin
Berlin, Germany

ISBN 978-3-030-10793-2 ISBN 978-3-030-10794-9 (eBook)
https://doi.org/10.1007/978-3-030-10794-9

Library of Congress Control Number: 2018968276

This Springer imprint is published by the registered company Springer Nature Switzerland AG
The registered company address is: Gewerbestrasse 11, 6330 Cham, Switzerland

Preface

This book is written for graduate students with a deeper interest in philosophy of science considerations concerning empirical marketing research. This interest may occur particularly at the beginning of an empirical research project that is typically part of a research paper or a dissertation. The book is most beneficial for this target group if the reader is familiar with basic methods of empirical research and statistical data analysis in the social sciences.

For empirical papers or dissertations, it is necessary to develop some relevant theoretical ideas as a foundation or—at least—to refer to existing theory. Therefore, this book focuses on the interface between philosophy of science aspects of theory development and the foundations of empirical methods. While the book covers basic concepts and ideas of social science methodology, it does not explain all technical details that are typically covered by social research or marketing research textbooks.

The book is partly based on some former publications of both authors on research methodology and marketing theory that were written in German language. If material from these publications was used in this book, it was thoroughly revised and complemented by numerous additional ideas and topics. The book is organized into ten chapters that cover philosophical and methodological considerations relevant for conducting a research project. Many examples and citations from leading authors ought to help readers to understand the content of this textbook as easy and clear as possible. For a deeper understanding, we suggested "further readings" at the end of each chapter and included many references in the text.

Both authors were (directly and indirectly) influenced in their thinking and work by Jack Jacoby. It was very sad to learn that Jack passed away in March 2018. He was a wonderful mentor, very inspiring with his clear and convincing thinking and his passion for research and teaching. He was not only a scientist "comme il faut" but a warmhearted person as well. Therefore, this book is dedicated to the memory of

Jack Jacoby

Simeon Todorov did an excellent job in preparing the English translation of the German manuscript and we are indebted to his support.

Frankfurt/Oder, Germany — Martin Eisend
Berlin, Germany — Alfred Kuss

Contents

Introduction

1

1.1 Characterization and Demarcation of Science

1.1.1 Essential Characteristics of Science

The concept of science is associated with considerable appeal and authority. Technological and material progress in many countries is largely due to the scientific developments of past centuries and decades, although this does not necessarily mean that every use of scientific knowledge is considered as progress. The field of science has been extremely successful over a long period of time and has gained a prestigious reputation. In some worldviews (for example, in Marxism), efforts are made to give authority to their particular viewpoint by establishing it as a scientific one, as "true" or "objective". In the field of politics or management, too, we can occasionally observe the practice of clarifying disputes through scientific investigations, which are supposed to bring about a "true" result. Sometimes, however, such investigations are designed, or manipulated, in such a way that the result that is desired by one of the involved parties eventually emerges.

What characterizes science and scientific research? This is a question that philosophers, among others, have been dealing with for many years. Here is a definition to begin with: "Scientific research is systematic, controlled, empirical, amoral, public, and critical investigation of natural phenomena. It is guided by theory and hypotheses about the presumed relations among such phenomena." (Kerlinger and Lee 2000, p. 14). The history of science analysis by David Lindberg (1992) provides a more detailed approach to the topic. Lindberg has put together some features, which are often regarded as typical of "science":

- *Science is an activity*: Science is seen here as an activity of people, which has led to an increasing understanding and control of the environment (in a broader sense).

M. Eisend, A. Kuss, *Research Methodology in Marketing*,
https://doi.org/10.1007/978-3-030-10794-9_1

- *Science is theory-oriented*: Science is based on theoretical knowledge, that is, an ordered set of concepts, facts and their relationships. Lindberg distinguishes science from technologies that involve the use of theoretical knowledge to solve practical problems.
- *Science searches for laws*: Science tries to discover general laws as precisely as possible. These can be used to explain real phenomena (see Sect. 2.3).
- *Science has specific methods of knowledge acquisition*: Scientific statements are developed and justified in a specific way, which is typically determined by logic, critical reflection and empirical verification.
- *Science is focused on specific topics*: Science refers to specific subject areas, e.g., "Marketing is specifically concerned with how transactions are created, stimulated, facilitated, and valued." (Kotler 1972, p. 49).

Bertrand Russell (1946, p. 549) concisely summarizes a central aspect of scientific work:

"It is not what the man of science believes that distinguishes him, but how and why he believes it. His beliefs are tentative, not dogmatic; they are based on evidence, not on authority or intuition."

The above considerations show how difficult it is to characterize or define the complex phenomenon of "science" in a single short sentence. In the formulation of three essential characteristics of science, Hans Poser (2001, pp. 21–22) refers to a definition by Immanuel Kant (1724–1804) ("If a doctrine is a system—i.e., a knowledge-total ordered according to principles—then it's what we call a science").

The three essential characteristics are:

- Science is first and foremost concerned with knowledge (see Sect. 1.2).
- Scientific statements must be justified.
- Scientific statements must form a system with an argumentative structure.

Another statement by Poser (2001, p. 11) adds another essential feature and an important requirement of science: **"Science manages. . .the best assured knowledge of its time."** This formulation also expresses the fact that scientific knowledge is related to a certain temporal context. In what follows, it will often be argued that knowledge is subject to errors or mistakes ("Fallibilism", see Sect. 1.2) and can later be replaced by "better" knowledge. Thus, for example, the medieval knowledge that the earth is the center of the universe was later replaced by better knowledge.

Alan Chalmers (2013, p. XX) conveys a vivid impression of the nature of science and emphasizes it in this way: "The fact that questions concerning the distinctiveness of scientific knowledge, as opposed to other kinds of

(continued)

knowledge, and the exact identification of the scientific method are seen as fundamentally important and consequential. As we shall see, however, answering these questions is by no means straightforward. A fair attempt to capture widespread intuitions about the answers to them is encapsulated, perhaps, in the idea that what is so special about science is that it is derived from the facts, rather than being based on personal opinion. This maybe captures the idea that, whereas personal opinions may differ over the relative merits of the novels of Charles Dickens and D. H. Lawrence, there is no room for such variation of opinions of Galileo's and Einstein's theories of relativity. It is the facts that are presumed to determine the superiority of Einstein's innovations over previous views on relativity, and anyone who fails to appreciate this is simply wrong."

Shelby Hunt provides a more pragmatic characterization of science (2010, pp. 19ff.) that comprises the following three characteristics:

- Science must refer to a "distinct subject matter" (also: object area), and have a specific common topic. In marketing, this is known as "exchange processes".
- The prerequisite for the (meaningful) application of science is the assumption of *similarities and regularities* with regard to the phenomena that constitute the object of science. In marketing, this applies to the effects of marketing measures (e.g., price elasticity) and to customer responses (e.g., development of customer relationships). If such regularities were not assumed, research would be meaningless because its results would not be applicable to comparable situations/ phenomena.
- On the basis of such regularities, science tries to formulate *laws, lawlike generalizations* (see Sect. 2.3.1) and to develop *theories* (see Chap. 2) in order to explain and predict the phenomena of interest. In marketing, for example, we want to understand the value of certain product features, how advertising works, or how different reward and payment systems affect the motivation of salespeople. Based on this understanding, we can develop and implement measures that lead to the desired effects.

For a more general characterization of science, the principles developed by the American sociologist, Robert K. Merton (1973, pp. 267–268), are relevant. These principles were presented in a book with the noteworthy title, "Science and Technology in a Democratic Order" (first edition 1942). The principle of "*universalism*" implies that all qualified scientists should contribute to scientific progress, without suffering from any ethnic, national, religious or other discrimination. The impact of discrimination, which is not only extremely painful for science, was experienced in Germany after the emigration and deaths of numerous scholars, in the years 1933–1945, due to their Jewish origin. Another principle states that scientific knowledge should be *available to the general public*. Merton (1973, p. 273) uses the concept of "communism", which is somewhat misleading today, and refers to the

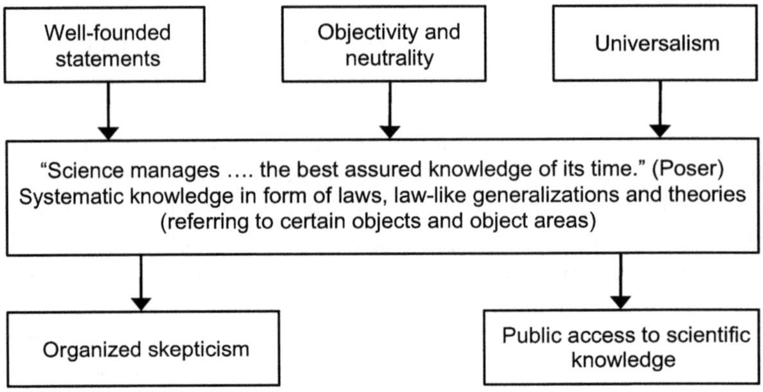

Fig. 1.1 Principles of science

shared possession of knowledge. "Secrecy" of relevant research results would significantly impede scientific progress and the (practical) use of science. The third point is the "*neutrality*" of scientists. Personal interests (for example, of a financial nature) should not influence scientific judgments. Threats to neutrality may occasionally arise in relation to third party funded research. Ultimately, "*organized skepticism*" is regarded as an essential principle of science. This means that the assessment of scientific statements should be free of political, religious or social influence, and should only be carried out according to the criteria of scientific methodology and critical review.

Figure 1.1 summarizes some essential aspects of the characterization of science, with the focus on the key features of science. The top line of Fig. 1.1 contains "input factors" of science. These are individual and well-founded statements that usually refer to existing knowledge, by means of logical derivation or by empirical data. In addition, the pursuit of neutrality and objectivity shapes the process of knowledge acquisition. Ultimately, the claim of universalism states that all qualified scientists can contribute to the process of knowledge generation and should be not subjected to personal, ethnic, religious, etc. discrimination.

As an "output" (bottom line in Fig. 1.1), it is noted that scientific knowledge should be available to the public; however, there are exceptions in practice, e.g., in the military sector or in industrial research. The term, "organized skepticism" means that (fallible!) scientific findings should always be subject to critical review. This starts with the review process for publications and ends with replacement through better knowledge by means of research.

In the literature it has been discussed whether science can or should be neutral. In this book, the **value neutrality** of science is explained by criteria introduced by Schurz (2014, pp. 37ff.). This position refers to the identification of research areas by Hans Reichenbach (1891–1953), with a distinction between the context of discovery and the context of justification.

The **context of discovery** encompasses questions about the origin of scientific statements and theories. Such questions are the subject of the fourth chapter of this book. However, there are important voices (especially Popper 2002, pp. 7ff.) that tend to view the creation of theories as a psychological problem and suggest that the scope of science should be focused on the justification and verification of scientific statements. But the emergence of hypotheses and theories might actually be relevant for research practice because it shows ways to gain new insights. For advanced and doctoral students, who are the target group of this textbook, it might be of interest to gain information on theory formation.

The **context of justification** refers to the examination of hypotheses and theories. The typical approach here is to derive (relatively concrete) hypotheses from existing theory, followed by empirical tests (see Chap. 5).

It should be noted that this distinction—apart from the above-mentioned criticism by Popper—is not undisputed. The criticism relates to the fact that the instruments for scientific discovery are still not very well developed and that the demarcation between both contexts is sometimes quite blurred (see, for example, Nickles 2008; Shapere 2000).

Later on, the **context of application** was added to the two categories mentioned above. This context refers to the use and application of adequately proven and/or reliable scientific knowledge for technical, medical, economic or social applications.

One can thus imagine a sequence of the emergence of scientific questions and hypotheses (context of discovery) to the systematic examination of hypotheses and theories (context of justification) up to the practical use of verified and tested knowledge (context of application). For example, one could raise the question of factors that influence sales force motivation, develop corresponding hypotheses and systematically test them using appropriate methods, and then apply and utilize this knowledge within the context of a company's sales force compensation policy.

In such a sequence Schurz (2014, pp. 37ff.) establishes his characterization of **value neutrality of science**. The first step (discovery) is, in many cases, influenced either by internal scientific values (for example, "Looking for yet unexplained phenomena or theoretical gaps"), as well as by external values (e.g., "Which research will receive sufficient funding?"). In the third step (application), both types of values play a role: from an internal point of view, scientists ask for the relevance of research results for further development of theory or for other areas of research. From an external value point of view, the purposes of the use of research results are critical, (e.g., gaining competitive advantage, increasing consumer satisfaction, military use), and of course also depend on the accessibility (secrecy vs. publication) and on the usage allowances (→ patents). The claim of value neutrality formulated by Schurz (2014, p. 42) relates only to the context of justification: "A specific realm of scientific activity, namely their context of justification, should be free from fundamental science-external value assumptions". This is the context of the main competences (and tasks) of science, namely the theoretical justification of statements, and their methodically appropriate (empirical) examination. Fig. 1.2 illustrates this approach.

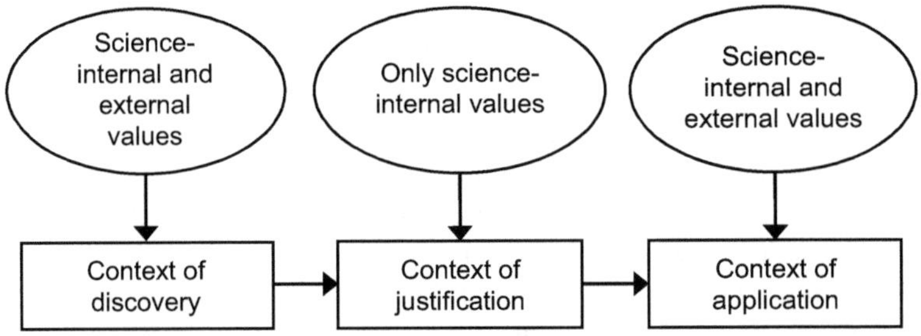

Fig. 1.2 Illustration of the requirement of value neutrality of science (following Schurz 2014, p. 43)

1.1.2 Science and Pseudo-science

To understand the nature of science it is helpful to delineate it from other—only seemingly scientific—activities, in particular so-called pseudo-science: "Pseudoscience is not science, it masquerades as science" (Monton 2014, p. 469). How can **science be distinguished from "pseudo-science"** (demarcation problem)? This is a question that has played an important part in philosophy of science discussions for decades. James Ladyman (2002, p. 265) defines the **demarcation problem** as follows: "The problem of providing a general rule or criterion for distinguishing science from non-science, and especially for distinguishing genuine science from activities or theories that are claimed to be scientific but which are not."

The beginning of this chapter has already suggested that the particular authority of science can sometimes lead to the fact that certain interested parties try to abuse this authority. As mentioned above, some ideological groups want to present their views as "scientifically founded". This can sometimes be due to economic interests. Take, for example, companies that are trying to "scientifically" prove the technical superiority of their products, for instance, the efficacy of new drugs, etc. There are institutions financed by interest groups, with the main intention of serving the goals of the funding organizations through expert assessments, studies, etc. There are other areas where the boundaries are more fluid, due to the selective use of scientific knowledge to enhance own interest. Many applications of scientific research show the crucial importance of the distinction between science and pseudo-science (sometimes called "non-science"). Here are some examples (see Hansson 2014):

Medicine For providers of medical services and products, for regulating authorities, for insurance companies and above all for patients, it is important to know which therapies are based on scientific research and which are not, e.g., some may be generated by charlatans and could even result in patients' lives being endangered.

Jurisdiction Numerous court decisions use scientific opinions and expert hearings. Only when the most valid and up-to-date scientific knowledge is used, can decisions be just and be accepted (see, for example, Jacoby 2013).

Politics/Society The results of PISA studies (*P*rogram for *I*nternational *S*tudent *A*ssessment) stand in the OECD (*O*rganization for *E*conomic *C*o-operation and *D*evelopment) as an important and generally accepted measure for the success of pedagogical concepts and educational policy. If the achieved results were not scientifically well founded and comprehensible, how could these results be accepted?

Gerhard Schurz (2014, pp. 2–3) explains a recent example of the demarcation problem (the problem of the demarcation between science and pseudo-science). It is about the so-called "Creationism" theory oriented around Biblical creation history, which some religious groups in the US want to establish as scientific truth and install in the curricula of American schools in place of the Darwinian theory of evolution:

"The *demarcation* problem is highly significant in society. In this context, it consists of the question of which of our ideas have a claim to the status of objective scientific knowledge that should be taught in public educational institutions, as opposed to subjective opinion, political values, ideologies, or religious convictions. This question became politically explosive in the controversy surrounding the creationism. For example, Judge W.R. Overton in 1981 (….), and Judge Jones in 2005 (….) based their decisions against the teaching of creationism in school on demarcation criteria between science and religious belief proposed by philosophers of science (….)."

The above examples illustrate the relevance of the demarcation of science from other methods of opinion formation with regard to a wide range of applications. Nevertheless, it is of course also important, within science, to be able to assess whether statements on which further research is built fulfill the requirements of scientific knowledge. If this were not the case, any succeeding scientific result would be based on weak foundations.

Karl Popper's efforts to find a solution to the demarcation problem were particularly important. The background to this was Popper's confrontation with the Marxist theory of history and Freudian psychoanalysis, where it seemed that "empirical support" was always found, because the followers of these worldviews could explain virtually everything *afterwards* (Popper 1963, pp. 45ff.). Such experiences led Popper to find a criterion for the demarcation between science and non-science in his **falsification approach** (for some aspects of this approach, see Sect. 3.1). Popper (2002, p. 18) briefly summarizes the central idea: "I shall not require of a scientific system that it shall be capable of being singled out, once and for all, in a positive sense; but I shall require that its logical form shall be such that it can be singled out, by means of empirical tests, in a negative sense: *it must be possible for an empirical scientific system to be refuted by experience*". Therefore, the focus is on the falsifiability of statements by empirical evidence, not their verification.

"Falsifiability" is the *possibility* that a statement can be falsified. However, it is also the case that predictions, e.g., in astrology (for example, "In the coming month you will be lucky in love"), which are not likely to seen as scientific (see, for example, Thagard 1998), can easily be falsified. To this extent, falsifiability would be a necessary, but not a sufficient, condition of scientific statements. Furthermore, it is argued by some authors that not only is falsifiability required for the delineation of science, but also that *falsification attempts* are actually undertaken, and that negative results have an *impact* on particular theories (either acceptance or rejection of the theory) (see Hansson 2014). A behavior which furthers the acceptance or rejection of a theory, dependent on the result of confirming or repudiating observations, would distinguish scientists from ideologues, religious fanatics, etc., whose ideas are not influenced by facts. Figure 1.3 illustrates these criteria.

In their review articles, Sven Ove Hansson (2014) and Bradley Monton (2014) summarize some *common features* of pseudo-science:

- Orientation on people who are supposed to have high competence regardless of the veracity of their statements.
- Reference to few examples, not to systematic empirical research with reproducible results.
- Ignoring or denying contradictory facts.
- Efforts to immunize statements against falsification.
- No integration into the context of established scientific knowledge.
- Shifting the "need for evidence" to skeptics, not to supporters of statements.
- Imitation of a "scientific sounding" jargon.

The demarcation between science and pseudo-science is of considerable relevance, because scientific statements enjoy particular credibility and acceptance (see above). "The knowledge-claims of sciences have greater epistemological warrant than the knowledge-claims of the non-sciences" (Hunt 2010, S. 260). Against this background, the following section outlines some aspects of epistemology, that is, the question of the origin and nature of scientific knowledge.

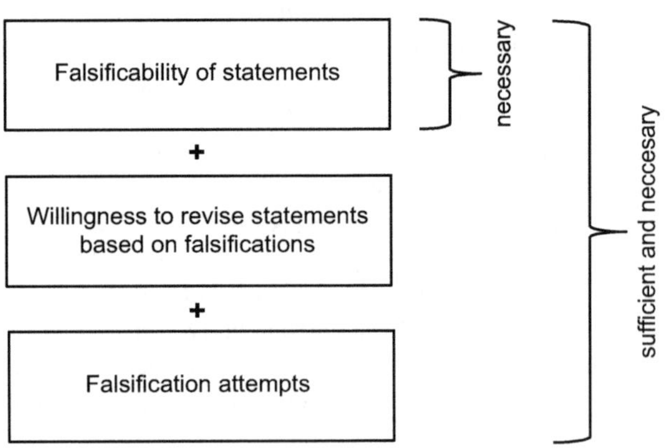

Fig. 1.3 Necessary and sufficient criteria for the demarcation between science and pseudo-science, following Hansson (2014)

1.2 Epistemology of Modern Science

The following chapters present and discuss in detail problems of the philosophy of science and their relevance for empirical research. In particular, Chap. 3 comprehensively describes the position of scientific realism that is central to this book. To begin with, an overview of the essential epistemological assumptions of the factual sciences gives a first impression of the role of empirical research in science. "In the term *factual sciences* (...), all sciences that have a part of the *real* world as area of enquiry are brought together; so, factual sciences include all groups apart from the formal sciences" (Schurz 2014, p. 29). This overview follows Gerhard Schurz's presentation of five central assumptions. In doing so, literal citations referring to this source are only indicated by the corresponding page number. The depiction of Schurz (2014) is compatible with the position of scientific realism (see Chap. 3), but not identical. Sometimes references to other parts of this book are given.

Epistemology is a part of philosophy and deals with the origin, limitations, and core essentials of knowledge. Schurz formulates a supreme epistemic goal and five epistemological assumptions, relevant for factual sciences. The restriction to factual sciences such as physics, chemistry, psychology and marketing is important, as in more formalistic sciences such as mathematics and logic, references to a particular reality and demands for empirical verification are, of course, of little relevance.

James Ladyman (2002, p. 5) briefly characterizes the central questions of epistemology:

"The branch of philosophy that inquires into knowledge and justification is called *epistemology*. The central questions of epistemology include: What is knowledge as opposed to mere belief? Can we be sure that we have any knowledge?; What things do we in fact know?. The first of these is perhaps the most fundamental epistemological question. Each of us has many beliefs, some true and some false. If I believe something that is, as a matter of fact, false (suppose, for example, that I believe that the capital city of Australia is Sydney) then I cannot be said to know it."

"The *supreme epistemic goal* (...) of science is to find *true* and *content rich* statements, laws, or theories, relating to a given domain of phenomena."(Schurz 2014, p. 19). This goal is closely linked to the considerations on the characterization of science in Sect. 1.1. The connection between truth and (information) content plays an essential role here. It is easy to make true statements with low content (e.g., "the market share of products can go up or down") or to formulate substantial statements of very doubtful truth (e.g., "the larger the number of employees, the greater the market success"). Scientists require special competence to arrive at statements that are simultaneously true and substantial. Theories play a central role because they are probably the most important way of summarizing and portraying scientific knowledge (see Chap. 2).

Schurz's first assumption may be somewhat surprising if one does not know about the fierce philosophy of science debates in the past decades. He calls this assumption "**minimal realism**" (p. 22), referring to a reality that exists independently of perceptions and interpretations of observers. Many people consider this self-evident. However, "minimal realism" is particularly emphasized here, because it is the counterview to relativistic and constructivist approaches (see Sect. 3.1), which played a certain role in the social sciences in the last third of the twentieth century, assuming that scientific theories are by no means an independent reflection of a reality, but are strongly context-dependent. According to the above-mentioned ultimate goal of science (as formulated by Schurz, among others), science exists, as far as possible, to deliver true and meaningful statements. Here "truth" is understood as the correspondence between these statements and the parts of reality under consideration ("correspondence theory of truth", see Sect. 2.2).

Against the background of many experiences from the history of science, the second assumption is called "**fallibilism and critical attitude**" (p. 23). This refers to the failure of scientific statements, even if at a certain time these statements appeared to be convincing and empirically supported. Therefore, (absolute) assurance as to the truth of scientific statements does not exist. This results in the "critical attitude" as the questioning of existing knowledge and the search for "better" statements and theories (see Sect. 3.1).

The public's predominant idea of science corresponds with the third assumption of "**objectivity and intersubjectivity**" (p. 23). This satisfies the already mentioned idea (epistemic goal), that a reality exists independent of the perception of the observer and that (in the ideal case) true statements are to be made about it. In this perspective, scientific statements should be as free as possible of personal and social interests, perceptions or goals. However, considerable objections exist to the possibility of arriving at completely objective statements, and will be discussed in Chap. 3. Nevertheless, if one were to abandon the claim of objectivity and intersubjectivity, one would not be able to meet the goal of approaching truth with scientific statements.

Julian Reiss and Jan Sprenger (2014) characterize scientific objectivity in the following way:

"Scientific objectivity is a characteristic of scientific claims, methods and results. It expresses the idea that the claims, methods and results of science are not, or should not be influenced by particular perspectives, value commitments, community bias or personal interests, to name a few relevant factors. Objectivity is often considered as an ideal for scientific inquiry, as a good reason for valuing scientific knowledge, and as the basis of the authority of science in society. (.....)

The ideal of objectivity has been criticized repeatedly in philosophy of science, questioning both its value and its attainability."

Accepting the fourth assumption of "**minimal empiricism**" (p. 23) expresses the fact that empirical research is an indispensable part of so-called factual science. This aspect is central to the following chapters because the relationship of theory development and testing with empirical methods is the main subject of this book.

What is meant by the term "**empirical**"? Generally, it refers to the idea of scientific knowledge being based on real-world experience or observations. Above all, we are concerned with the examination of theoretical statements with regard to the extent to which they correspond to reality. In empirical marketing research, this is usually done in such a way that theoretically founded hypotheses are formulated, from which prognoses for observations in reality are deduced (assuming the validity of the theory), and by corresponding results of these observations the theory is confirmed (see Chap. 5). Another practicable way of assessing a theory is to depict it (or parts of it) as a model, and to determine to what extent the model is consistent with real data (see Chap. 7). "Over the long run, scientific conceptualizations tend to be accepted (at least tentatively) only to the extent that they have been subjected to rigorous, systematic empirical testing and have been demonstrated, via replication, to hold up and prove useful in understanding reality" (Jacoby 2013, p. 187).

In the field of marketing, empirical research has a long tradition and is still of great importance. On the one hand, this is probably due to the fact that, since the 1970s, a behavioral-scientific orientation has developed with a strong empirical orientation, which is not surprising in research areas such as "advertising" or "consumer behavior". On the other hand, market research, which provides the methodological tools for empirical research, has been firmly established in the textbooks and curricula of marketing since the 1950s. A strong need for empirical research, combined with training of marketing researchers in empirical methods, explains the strong empirical orientation in marketing research. Meanwhile, there is literature (Yadav 2010) that criticizes the dominant proportion of empirical research and a small and shrinking number of conceptual works. However, it must be borne in mind that in the marketing literature, new theoretical approaches are often published only when they are already based on some empirical evidence. In any case, marketing researchers should be familiar with the methods of empirical research, in order to develop a critical understanding of the literature. The goal of the present book is also to contribute to this goal.

Schurz refers to "**logic in the wider sense**" (p. 24) as the fifth assumption. This primarily means that concepts are precisely defined in theories and that statements must be precisely formulated so that the corresponding meanings are determined. Otherwise (in blurred terms and statements), it would not be possible to determine the truth content of a theory and one could not formulate sufficiently accurate hypotheses. For instance, how to establish and verify a hypothesis about the link between the amount of information and the decision-making quality, if one does not specify exactly what "decision-making quality" really means (for example, acceptance or speed or financial consequence of a decision)?

Figure 1.4 and the subsequent brief explanations summarize and illustrate the essential aspects and contexts of the five epistemological assumptions formulated by Schurz (2014, pp. 19ff.). Supporting the background of the central theme of this

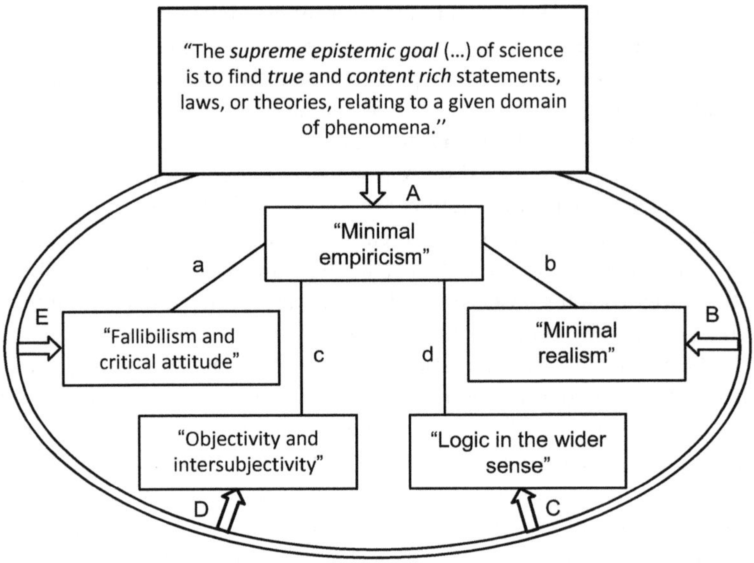

Fig. 1.4 Supreme epistemological goal and epistemological assumptions (following Schurz 2014, pp. 19ff.) (Direct quotes by Schurz 2014)

book, the relationship of "minimal empiricism" to other assumptions will be discussed a little more extensively.

Figure 1.4 briefly explains the various relationships. First, we will look at the relationships with the *supreme epistemic goal* (in particular, the aspects of the alleged truth of statements, laws, and theories) with the five epistemological assumptions:

(A) Factual sciences (i.e., sciences that make statements about certain parts of reality such as consumers, media, and services) can gain and verify such statements by empirical methods. "Empirical observations (. . . .) are a decisive referee in the scientific search for truth: scientific law hypotheses and theories must be *tested* in terms of them" (Schurz 2014, pp. 23–24).

(B) There is a reality that is independent of the perceptions and interpretations of observers (see Psillos 2006). The truth of scientific statements is determined by the degree of correspondence between a (theoretical) statement and the corresponding part of reality. Without realism, "truth" in this sense would not be possible.

(C) Only precise definitions, statements and arguments allow precise comparisons between theory and the real world, and thus assessments of the truth of statements, laws, and theories.

(D) "A statement's truth must hold objectively, i.e., independently of the beliefs and value-attitudes of the epistemological subject, because by assumption (minimal realism), reality has subject-independent existence, and truth is the correspondence between statement and reality" (Schurz 2014, p. 23).

(E) Logic and historical experiences teach us that one can never be sure of the truth of statements. One can, however, try to reach statements that are *approximately true* (see Sect. 3.1). A critical attitude repeatedly questions the truth content of statements and further increases it via appropriate research.

In regard to empirical research, the following relationships with the abovementioned assumptions play a crucial role:

(a) Empirical research is the central tool for examining how truthful existing knowledge is and to determine whether alternative/new approaches lead to greater proximity to the truth.
(b) Empirical research provides the methods that, in many cases, make it possible to make real-world observations, to analyze them and to determine the degree of correspondence between theory and reality.
(c) Empirical results on the truthfulness of statements can only be convincing if they are not influenced by measurement errors and subjectivity. For this reason, the validity of research results (see Chap. 6) must be ensured by careful application of adequate methods and verified by means of appropriate documentation in publications.
(d) "Logic in the wider sense" refers to the precision of statements. Without such precision, it is not possible to develop adequate measurement instruments (→ validity) and to determine the extent to which theoretical hypotheses and real observations correspond. How, for example, should the suitability of a measurement scale for "brand loyalty" be assessed, if it is has not been precisely defined what "brand loyalty" is?

Gerhard Schurz (2014, p. 26) supplements his epistemological assumptions (see above) by some "common methodological features" (M1 to M4). He uses the term "actual observation sentences", which means sentences whose content corresponds to actual observations.

"M1: Science searches for hypotheses which are as general and as content rich as possible, and recorded in a scientific language. In all disciplines, these hypotheses include ***laws*** and ***theories*** (. . ..).

M2: Science searches for ***actual observation sentences***, as many as possible (and as relevant as possible), which reflect the results of observations, experiments or measurements.

(continued)

M3: Science attempts (.....) to ***explain*** the currently known actual observation sentences, and to predict new, and yet unknown ***potential*** observation sentences.

M4: Science attempts to ***test empirically*** its general and hypothetical sentences by comparing the predicted (potential) observation sentences with the currently known (actual) observation sentences. If the latter are in agreement with the former, the prediction was successful (it then becomes a successful explanation) and the hypothetical sentence (law or theory) is ***confirmed*** or ***corroborated***. If there is a contradiction between the latter and the former, the prediction was without success and the law or theory is ***falsified***, or in the case of a merely statistical prediction ***weakened***".

The basic idea in M3 will be discussed again in the section on explanations and predictions (Sect. 2.4). A more concrete development of the idea presented in M4 is discussed in connection with the hypothetical-deductive model (Sect. 5.2) and the inductive-realistic model (Sect. 5.3).

This book focuses on empirical marketing research. What contributions does marketing research provide from an epistemological point of view? So far, the view has mainly focused on the application of empirical methods for *testing theories*—both existing and newly developed ones—by means of formulating hypotheses and their empirical examination. Figure 1.5 illustrates this procedure. In addition, ("classical") applications of empirical research for *theory building* (see Chap. 4) are outlined below. In addition, research concerning *empirical methods* (see the end of this section) plays a crucial role. Methods are important in multiple areas of marketing research because a large part of the phenomena of interest cannot be examined by mere eyeballing or using generally accessible information. It is often necessary to develop particular ways to measure, for instance, customer satisfaction or willingness to pay.

Figure 1.5 shows important steps in the empirical testing of theories. At the beginning (of course), there is the theory to be tested, followed by hypotheses, i.e., expectations regarding certain expressions or relationships of characteristics in reality. To test the hypotheses, suitable measurements are required, which are often already known and available in the appropriate research field (e.g., experiments and surveys). The application of these methods leads to results, which allow conclusions in regard to the degree of confirmation of the tested theory.

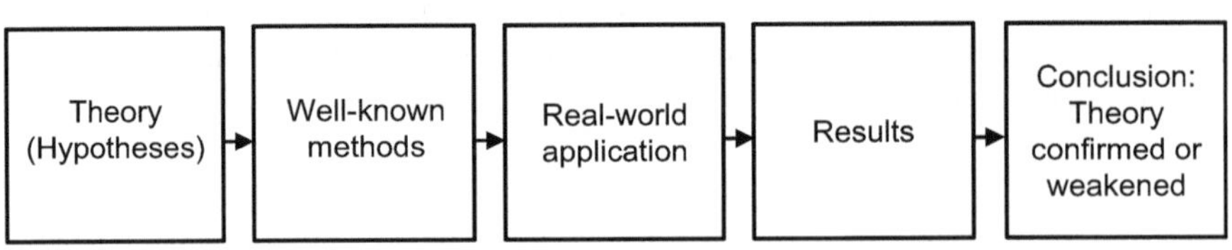

Fig. 1.5 Illustration of empirical research for testing theories

Figure 1.6 investigates two further applications of empirical research: theory building and interpretation of applied research (see Chap. 4).

The sequence shown in Fig. 1.6 thus refers to two different applications. On the one hand, *theory building* is depicted. If a research problem finds no (satisfactory) theory (e.g., the formation of credibility judgments with regard to product information offered on the Internet), then it is common (but by no means exclusive) practice to address this question with observations based on commonly applied methods (e.g., exploratory interviews) and to develop a theory on the basis of these results (see the discussion on Grounded Theory in Chap. 4). On the other hand, for *applied questions* a corresponding approach (e.g., the definition of certain market segments) is often used. Typically, a theory is not of interest here. In these cases, descriptive data are often used (e.g., "How high was the share of new customers in the last year?") and the interpretation of connections between certain characteristics follows the empirical investigation (e.g., "In which regions was the share of new customers particularly high?"). In many cases, the findings provide practical implications and advice for actions (e.g., "strengthen customer acquisition").

Figure 1.7 portrays an entirely different type of research that does not refer to substantive questions (theory testing or building, practical application of results), but the *development of new* data collection and analysis *methods*.

Both in basic and applied research one needs measures for novel phenomena or one requires more powerful (analytical) methods. One example is the current development in the research of Internet use. In marketing research, one problem is that measurement instruments (e.g., certain survey techniques) are very error-sensitive and therefore a special process is required to develop valid measurement instruments (Churchill 1979). In view of a new methodological problem, a corresponding proposal is therefore developed and applied. Based on the findings, one can decide whether the new method is appropriate. There is a particularly well established practice for the development of measurement instruments, for which a

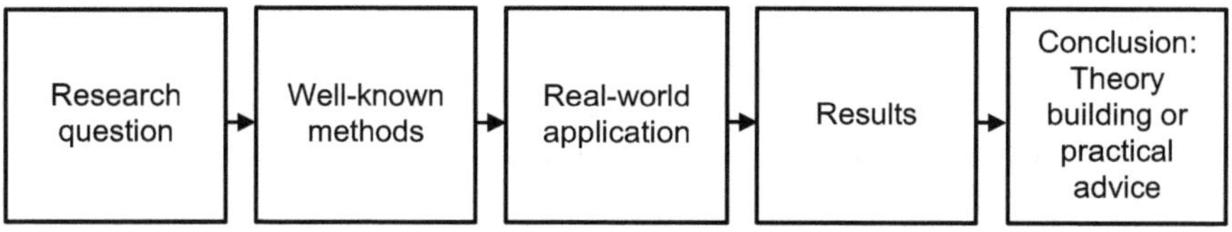

Fig. 1.6 Illustration of empirical research for theory building and applied research

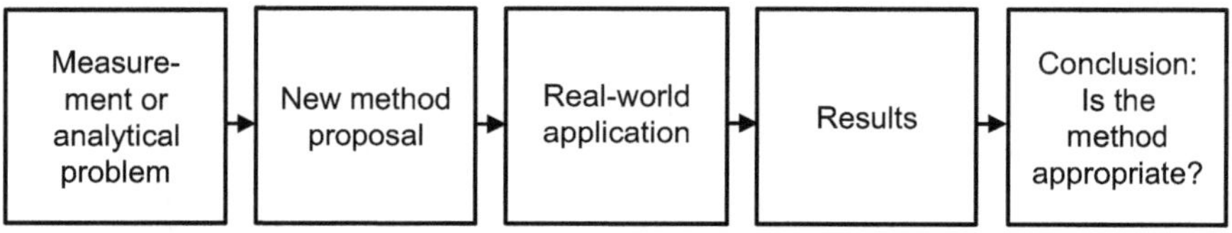

Fig. 1.7 Illustration of empirical research for method development

standardized approach is often used (see Churchill 1979; Rossiter 2002 and Chap. 6 of this book).

1.3 Contents and Structure of the Book

This section provides a brief overview of the contents and structure of the subsequent chapters. Theory building and testing is crucially important for generating scientific knowledge. The largest and most important parts of knowledge in various sciences are developed, preserved and passed on in the form of theories.

> Fred Kerlinger and Howard Lee (2000, p. 11) explain the role of scientific theories in the following way:
>
> "The basic aim of science is theory. Perhaps less cryptically, the basic aim of science is to explain natural phenomena. Such explanations are called 'theories'."

In light of this background, "the nature and relevance of theories" is presented and discussed in the second chapter. Empirical research, which is the main focus of this book, plays a decisive role in theory testing and theory building. In order to facilitate understanding, the book first explains the nature of theories, before discussing theory building and testing (see Fig. 1.8). Different epistemological positions play a significant role here. Since the middle of the twentieth century, critical rationalism (proposed mainly by Karl Popper), relativism and constructivism, and, more recently, scientific realism are important views. The literature discusses these positions extensively and comprehensively, but this book does not delve deeply into these discussions. Rather, the dominant position of scientific realism is assumed, explained and justified in Chap. 3. For the purpose of delineation, the other philosophy of science positions are briefly addressed. These epistemological considerations are presented before the discussion of theory building and testing, because both aspects are influenced by these considerations.

The reflections in Chap. 5 present the role of empirical research as a "referee" (Schurz 2014, p. 24) with regard to the suitability of theories for the understanding of reality.

Chaps. 6, 7 and 8 discuss the methodological aspects of the empirical testing of theories. In Chap. 6, basic questions of data collection (i.e., measurement and operationalization) are discussed. Chapter 7 deals with statistical tests and modeling. Tests of causal hypotheses, which have special importance for theory and practice, are the subject of Chap. 8. These chapters do not deal with methodological details, for which extensive literature exists. Rather, they are about basic scientific considerations on the application of these methods. Chaps. 6, 7 and 8 are mainly concerned with the design and interpretation of studies and their results for theory testing. Chapter 9 is concerned with the generalization of a larger number of

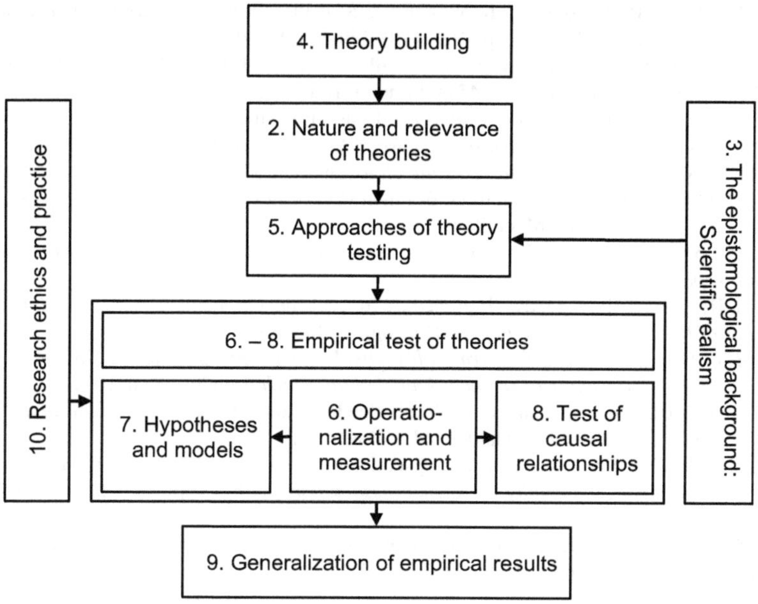

Fig. 1.8 Overview of content and structure of the following chapters

empirical results and their correspondence with theory. The end of the book ("last, but not least!") carries reflections on "research ethics and research practice", which have been given particular relevance by some scientific scandals in the recent past. In the meantime, the science system has developed regulations for the prevention of unethical behavior, which are also a prerequisite for successful research activities.

References

Chalmers, A. (2013). *What is this thing called science?* (4th ed.). Indianapolis, IN: Hackett.

Churchill, G. A. (1979). A paradigm for developing better measures of marketing constructs. *Journal of Marketing Research, 16*(February), 64–73.

Hansson, S. (2014). Science and pseudo-science. In E. Zalta (Ed.), *The Stanford encyclopedia of philosophy*. http://plato.stanford.edu

Hunt, S. D. (2010). *Marketing theory—Foundations, controversy, strategy, resource-advantage theory*. Armonk, NY: Sharpe.

Jacoby, J. (2013). *Trademark surveys—Designing, implementing, and evaluating surveys*. Chicago: American Bar Association.

Kerlinger, F., & Lee, H. (2000). *Foundations of behavioral research* (4th ed.). Melbourne: Wadsworth.

Kotler, P. (1972). A generic concept of marketing. *Journal of Marketing, 36*(April), 46–54.

Ladyman, J. (2002). *Understanding philosophy of science*. London: Routledge.

Lindberg, D. (1992). *The beginnings of Western science*. Chicago, IL: University of Chicago Press.

Merton, R. (1973). *The sociology of science—Theoretical and empirical investigations*. Chicago, IL: University of Chicago Press.

Monton, B. (2014). Pseudoscience. In M. Curd & S. Psillos (Eds.), *The Routledge companion to philosophy of science* (2nd ed., pp. 469–478). London: Routledge.

Nickles, T. (2008). Scientific discovery. In S. Psillos & M. Curd (Eds.), *The Routledge companion to philosophy of science* (pp. 442–451). London: Routledge.

Popper, K. (1963). *Conjectures and refutations*. London: Routledge.

Popper, K. (2002). *The logic of scientific discovery*. London: Routledge.

Poser, H. (2001). *Wissenschaftstheorie—Eine philosophische Einführung*. Stuttgart: Reclam.

Psillos, S. (2006). Scientific realism. In D. Borchert (Ed.), *Encyclopedia of philosophy* (Vol. 8, 2nd ed., pp. 688–694). Detroit: Macmillan.

Reiss, J., & Sprenger, J. (2014). Scientific objectivity. In E. Zalta (Ed.), *The Stanford encyclopedia of philosophy*. https://plato.stanford.edu

Rossiter, J. R. (2002). The C-OAR-SE procedure for scale development in marketing. *International Journal of Research in Marketing, 19*(December), 305–335.

Russell, B. (1946). *A history of Western philosophy*. London: George Allen & Unwin.

Schurz, G. (2014). *Philosophy of science. A unified approach*. London: Routledge.

Shapere, D. (2000). Scientific change. In W. H. Newton-Smith (Ed.), *A companion to the philosophy of science* (pp. 413–422). Malden, MA: Blackwell.

Thagard, P. (1998). Why astrology is a pseudoscience. In P. Asquith & I. Hacking (Eds.), *Proceedings of the philosophy of science association* (Vol. 1, pp. 223–234). East Lansing, MI: Norton.

Yadav, M. S. (2010). The decline of conceptual articles and implications for knowledge development. *Journal of Marketing, 74*(January), 1–19.

Further Reading

Hunt, S. D. (2010). *Marketing theory—Foundations, controversy, strategy, resource-advantage theory*. Armonk, NY: Sharpe.

Schurz, G. (2014). *Philosophy of science. A unified approach*. London: Routledge.

The Nature and Relevance of Theories 2

2.1 Basic Ideas

Theory and empiricism are closely connected. Empiricism, on the one hand, plays a central role in the confirmation of existing theories: positive results of empirical tests lead to increased acceptance of theories, while negative results tend to contribute to the rejection or modification of theories. On the other hand, certain empirical methods help to develop theories (see Sect. 4.3). Therefore, the description and discussion of theories must play a key role in a textbook on the fundamentals of empirical research. This chapter deals with the essential features of theories and their relevance. Chapters 4 and 5 then deal with aspects of theorizing and ways of testing and assessing theories. What does the term "theory" actually mean? Karl Popper (2002, p. 37 f.) gives a description that allows a first look at the work on and with theories: "Theories are nets cast to catch what we call 'the world': to rationalize, to explain, and to master it. We endeavor to make the mesh ever finer and finer."

Essentially, **theories** are linguistic entities (often also—partially—formulated in the language of mathematics and/or graphically illustrated) that formulate assertions that may be shown to be correct or false in a (later) test. As so often in science, the conceptions of the nature of theories are not quite consistent. However, it is possible to identify central features (which is attempted here) that are widely agreed in marketing research. Here are three definitions for the term "*theory*", which clarify the essential elements of theories:

- "A theory is a set of statements about the relationship(s) between two or more concepts or constructs" (Jaccard and Jacoby 2010, p. 28).
- "A theory is a set of interrelated constructs (concepts), definitions, and propositions that present a systematic view of phenomena by specifying relations among variables, with the purpose of explaining and predicting the phenomena" (Kerlinger and Lee 2000, p. 11).

M. Eisend, A. Kuss, *Research Methodology in Marketing*,
https://doi.org/10.1007/978-3-030-10794-9_2

- "A theory is a systematically related set of statements, including some lawlike generalizations, that is empirically testable" (Hunt 2010, p. 175).

Now to the interpretation of these different definitions (regarding the terms "concept / construct" see below): Kerlinger and Lee's formulation, as well as Hunt's, show that—at least in the field of the social sciences (including marketing)—theories are mental entities that are suitable for describing and explaining a multitude of corresponding *phenomena of reality*. So here it is about the identification of more general (i.e., beyond the individual case) regularities or *laws* (see Sect. 2.3.1). The cited authors also emphasize the aspect of *systematization*, that is, the organized summary of individual concepts, statements, etc., for an adequately comprehensive presentation of a part of reality. This implies that a theory is about a *set of statements*. Hardly anyone would call a single statement a theory (e.g., "As risk perception increases, the information demand increases"). Rather, the representation of a larger number of relationships between relevant phenomena (including cause-effect relationships) is characteristic of a theory. Theorizing typically lies between a sufficiently accurate (and thus often complex) representation of real phenomena, on the one hand, and the pursuit of simplicity and comprehensibility on the other hand (Hunt 2015). However, reviews ("tests") of theories can well be limited to a single or a few aspects of the theory and hypotheses (see Sect. 5.2).

Based on a (verbally communicated) proposal by Ajay Kohli, the essential **elements of a theory** are as follows:

- Concepts with the corresponding definitions
- Statements about relationships between the concepts
- Arguments that substantiate the statements.

The third aspect (arguments that justify a theory) plays a central role in the acceptance of a theory (e.g., in publications). This will be discussed related to a model of theory formation in Sect. 4.3.1. The definitions by Jaccard and Jacoby (2010) and by Kerlinger and Lee (2000) use the terms "Constructs / Concepts". These terms are synonymously used in this book as well. **Concepts** (and in the viewpoint represented here, also **constructs**) are abstractions (and therefore generalizations) of individual phenomena in reality, which are appropriate for the respective point of view. Hardly anybody is concerned, for example, with the huge variety of physiological processes in a human body, but most people speak—if there are no relevant problems—in an abstract and summarizing way of "health". Here is a second example from everyday life: in general, one does not deal with the differences between many entities with four wheels and a motor, but uses—for instance, when analyzing traffic flows or corresponding markets—the *concept* "car" (which abstracts from technical details and differences). Concepts serve to summarize a variety of objects or events in terms of common characteristics and deferring other differences. Thus, they simplify the reality and in this way become indispensable "building blocks of understanding" (Jaccard and Jacoby 2010, p. 10).

The process of creating concepts, the so-called **conceptualization**, will be discussed in Sect. 4.1.

Jaccard and Jacoby (2010, p. 11 ff.) fully explain concepts as outlined above, which are of fundamental importance to scientific work (including theory building):

- "*Concepts are generalized abstractions.*" A concept stands for a general idea under which a multiplicity of (different) forms of similar phenomena is summarized in the relevant perspective. While there are millions of different cars, the concept of "car" summarizes essential common features of cars. As such, certain details (e.g., color, brand, price) are abstracted. The delineation of such concepts is often not very easy or unambiguous. For instance, it is not trivial in marketing to clearly define concepts such as "advertising impact" or "product quality".
- "*Concepts encompass universes of possibilities.*" Following the above point of view, one can say that concepts encompass a spectrum of objects and phenomena that are different to some extent. For example, the concept of "car" includes items with four wheels, an engine, etc., but they may differ significantly in terms of various features (e.g., size, outer shape, top speed). If one uses the concept of "manager", the concept covers a wide variety of people in different industries, companies and functional areas that perform certain types of tasks.
- "*Concepts are hypothetical.*" This characteristic is quite obvious when one thinks of concepts such as "happiness" or "solidarity". But even in the example of cars, the abstraction process of conceptualization shows that it is no longer about individual concrete objects (such as the neighbor's white BMW), but rather about a comprehensive and thus abstracting view.
- "*(Most) concepts are learned.*" In the socialization process, we learn which concepts exist for which objects, situations, etc. For example, one needs such learning processes to understand concepts such as "compact" or "sensitive". When studying marketing we learn concepts such as "market segment" or "relative market share".
- "*Concepts are socially shared.*" One understands in English-speaking countries quite uniformly what the concept "car" means. Young people have (so it seems to the outside observer) a quite uniform understanding of the concept "cool".
- "*Concepts are reality oriented (or functional).*" Concepts have a function for the interpretation and understanding of reality. Without a corresponding understanding, for example, of the phenomenon "attitude", we cannot apply certain knowledge (e.g., "attitudes influence behavior").
- "*Concepts are selective constructions.*" Concepts depend on the perspective of interest. For example, one can assign the same person—depending on the perspective—to concepts (categories) such as woman, academic, jogger, opera lover, etc. In this respect, concepts can also be "theory-laden" because existing theoretical ideas and interests can influence or shape the perception of reality (see also Sect. 3.2).

Kotler and Keller (2012, p. 30) use the concept "market" as an example of the fact that a given concept can have quite different meanings in different groups or that the same concept can stand for different contents. This suggests that precise definitions of concepts (see Sect. 4.1) are extremely important for theory and empirical research.

"Traditionally, a 'market' was a physical place where buyers and sellers gathered to buy and sell goods. Economists describe a market as a collection of buyers and sellers who transact over a particular product or product class (such as the housing market or the grain market). (. . .) Marketers use the term market to cover various groupings of customers. They view sellers as constituting the industry and buyers as constituting the market."

By using concepts, theoretical understanding of the vast variety of real "objects" (e.g., organizations, people, characteristics) and their relationships becomes feasible. The communication of scientific statements needs (especially in the literature) designated concepts for corresponding **terms**. This assignment is not always easy or even clear in marketing research. Precise definitions (see Sect. 4.1) play a central role. The difference between terms and concepts is that terms are usually tied to a particular language. Thus, the terms "brand", "marca" (Spanish), or "Marke" (German) refer to the same real phenomenon; it is an identical concept, to which various terms are assigned.

The connection between concepts, terms and objects is explained by Bagozzi (1980, pp. 114–115):

"A concept may be defined as the basic unit of thought (. . .). It represents a mental construct or image of some object, thing, idea, or phenomenon. More formally, concepts achieve their meaning through their relationships with terms and objects (where objects are construed broadly to include physical things, events or happenings, etc.). As shown in the figure, it is possible to represent these relationships as connections among three systems of worlds of meaning. The relationship between a concept and term is one between the

(continued)

world of thought and the world of language."

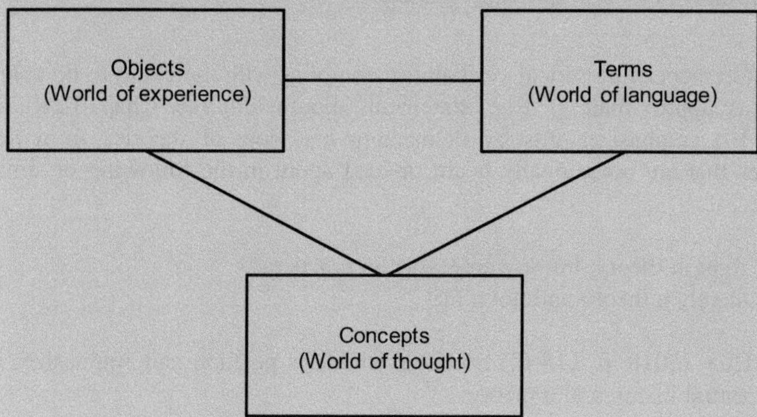

At the beginning of this chapter, theories were also characterized as making statements about *relationships between concepts*. These relationships can be specified in different ways and with varying degrees of precision. Thus, statements about relationships are not confined only to the assumption of a positive or negative relationship, they can also be more precise in terms of the nature of the relationship (e.g., linear or non-linear). The more detailed description of such relationships plays a role in the refinement of theories (see Sect. 9.4). A special kind of relationship—so-called *causal relationships*—is discussed in detail in Chap. 8. If one has theoretically grounded assumptions about relationships between concepts, then one can formulate corresponding **hypotheses** and test them empirically (see Chap. 5). What is meant by this? Hypotheses are theoretically founded assumptions about facts or relationships that have not yet been empirically confirmed. In many cases, hypotheses refer to relatively concrete relationships that will be tested with the intention of assessing the validity of a more general theory (Jaccard and Jacoby 2010, p. 76). This approach, in which hypotheses are derived ("deduced") and empirically tested from a theory, is called a **hypothetico-deductive method** (see Sect. 5.2). The corresponding tests in the social sciences are the subject (above all) of empirical research, as will be discussed in the chapters that follow. In this view, we may consider a hypothesis as a theoretically based prognosis for a particular empirical result (e.g., a positive linear relationship between the variables X and Y). Corresponding empirical research can then show whether the prognosis (hypothesis) is confirmed by the real data. This aspect will play a significant role in the context of the inductive-realistic model (see Sect. 5.2).

The idea of empirical assessments of theories is widely accepted (not only) in marketing research and Richard Rudner (1966, p. 10) includes it in his definition of social science theories:

"A theory is a systematically related set of statements, including some lawlike generalizations, that is empirically testable."

The requirement of empirical verifiability complies with the claim to be able to make (at least approximately) "true" statements about reality (see Chap. 3). Rudner (1966, p. 10) emphasizes this by delineating his view of theories from *false* perspectives that are occasionally heard or read about in the following or similar phrases:

- "It's all right in theory, but it won't work in practice."
- "That's merely a theory and not a fact."

Shelby Hunt (2010, p. 175 ff.) backs up Rudner's position and emphasizes the following central features of a theory:

- **Systematic relationships** between the statements contained in a theory. Systematization is associated with consistency of statements and allows an understanding that would not be possible in an unsystematic accumulation of (individual) statements.
- **Lawlike statements**, that is, statements about well-founded regularities (if-then relationships), which are independent of time and place. These allow explanations and predictions (see Sect. 2.4) of phenomena. Statics laws, for example, explain why a bridge can withstand a certain load and also allow a prediction of its load capacity if certain design features are known.
- **Empirical testability**, because tests of the correspondence of theory and reality show—at least from the perspective of scientific realism (see Chap. 3)—whether a theory is more or less true, regardless of the observer's views, wishes, or ideologies.

Richard Rudner (1966, p. 11) on the systematization of statements in a theory:

"We are all familiar with the view that it is not the business of science merely to collect unrelated, haphazard, disconnected bits of information; that it is an ideal of science to give an organized account of the universe—to connect, to fit together in relations of subsumption, the statements embodying the knowledge that has been acquired. Such organization is a necessary condition for the accomplishment of two of science's chief functions, explanation and prediction."

For the empirical verification of theories, the above-mentioned lawlike statements play an essential role, with which the following chapters will deal with respect to the hypothetico-deductive method. In the process, statements are derived from laws and the correspondence of these statements with real observations is the decisive criterion for the empirical validation of the respective theory. Schurz (2014, p. 28)

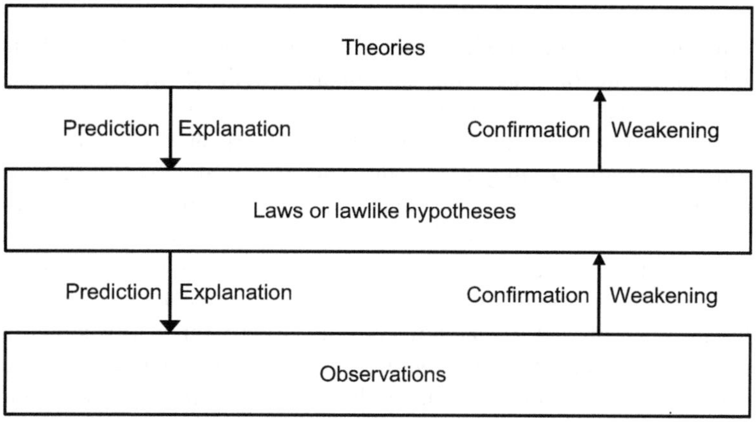

Fig. 2.1 Relationship between theories, laws, and observations (according to Schurz 2014, p. 28)

summarizes the relationship between theories, laws, and observations that is depicted in Fig. 2.1.

For example, one could derive from a more general theory of information processing a **law-based hypothesis** whereby, with some degree of information overload, the processing of information goes down and "peripheral stimuli" are preferably used to form an attitude (see the following example of the Elaboration Likelihood Model). If one goes from the theory to laws and to observations, then this can serve to explain a real observation through a corresponding law and the theory behind it. In this way one can come to the prediction of observations even if observations are yet to be made (assuming that the theory is approximately true). At the same time, this is also relevant for the theory test when observations are compared with the corresponding expectations on the basis of theories or laws; depending on the results, the respective theory or law is confirmed or weakened (in the case of the "inductive-realistic model" in Chap. 5, this is called "success" or "failure").

It has become clear that law-based hypotheses are an essential part of theories and also have a central function for their applications with regard to the explanation and prediction of real phenomena. The same applies to the testing of theories by empirical observations. Therefore, one finds numerous empirical studies that do not (relatively broadly) refer to *whole theories*, but are limited to the test of a few law-based hypotheses. Such examinations will be expressly included when referring to "theory testing" or "theory examination" in this book.

As an illustrative example of a theory that has received a great deal of attention in marketing and communication research, the Elaboration Likelihood Model (ELM) is outlined here. This model goes back to Richard Petty and John Cacioppo (see, for example, Petty et al. 1983) and has been reviewed and confirmed by them and by other authors in numerous studies.

(continued)

A key feature of the model is the distinction between a "central" and a "peripheral" route of processing information, both of which can lead to attitude changes. On the *central* route, intensive information processing takes place through evaluation of attributes, comparison of alternatives, comparison with already existing information, etc. The result of such a process can be a relatively stable attitude change, which is largely determined by the content and relevance of the information contained in a message (i.e., arguments). However, this route, which involves a great deal of processing, only applies if a person is appropriately motivated and capable of processing all this information. Only very few consumers are willing to study comprehensive information material (advertising leaflets, test reports, etc.) before purchasing relatively unimportant products (paper handkerchiefs, batteries, etc.). In many cases, the skills for understanding and processing the information are not available, for example, because of intellectual limitations or lack of specific expertise.

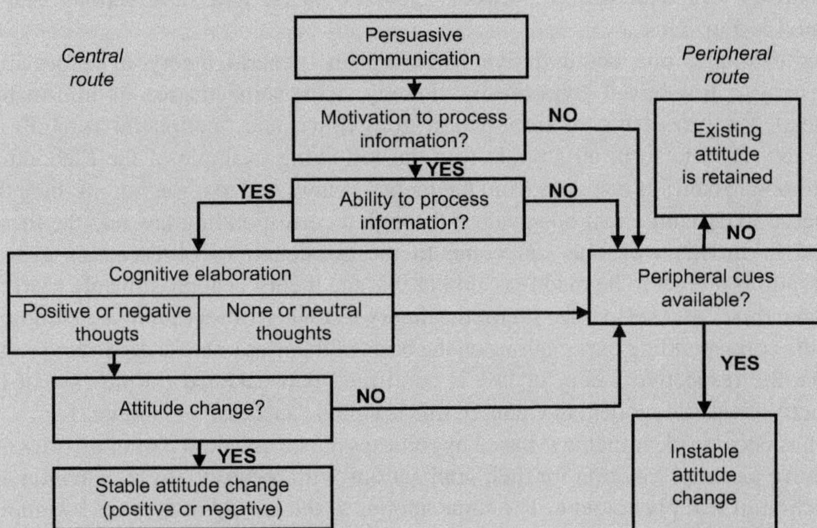

(continued)

The *peripheral* route of processing a message applies when the motivation and/or ability to comprehensively process the message is missing. Then, so-called "peripheral cues" play a major role. These are easily processed features of the message that have little to do with their content (e.g., the advertised product), such as the aesthetic design of an advertisement or the attractiveness of endorsers appearing in the message. Given such stimuli, they can result in a (relatively weak) and less stable attitude change. The illustration gives a corresponding overview.

This example illustrates the features of a theory as outlined above:

The **system of relations** of statements is already clearly recognizable from the illustration. One can clearly see which aspects relate to each other in which way.

The **generalizability of laws** becomes obvious by the terminology used in the ELM. It is all about "influencing communication", not just advertising or communication under very specific conditions. The other terms used (e.g., "motivation or ability to process information") also have a fairly high level of generality. On the other hand, a very specific statement (e.g., "The attitude of person X to brand Y is more influenced by an advertisement if person X sees it at breakfast and not in the subway") does not meet the demands on the generality of a theory.

The **empirical testability of the statements** is present throughout. The corresponding hypotheses are derived directly from the model and can be tested, for instance, by experiments. As for the ELM, this has already happened in numerous studies.

Even during the first year of studies a student can notice that theories play a central role in all sciences. The collection of knowledge and scientific progress leads the further development and modification of theories. Theories summarize the state of knowledge of a subject, and in the form of theories, this state of knowledge is available to a broader public, not least the users who apply this knowledge in practice. Knowledge transfer takes place through textbooks and other publications as well as through lectures at universities and similar institutions. In this sense, the **relevance of theories**—irrespective of subject area—is that they are the most important source of information for scientific knowledge.

Why are theories so important to science beyond the general aspect above? Why does science focus on theories, although for some "clients" of science—practitioners and students—the interest in theories is not always so obvious? What relevance do theories have? Here are some considerations and experiences:

- One can easily understand the importance of **order and structuring** if one imagines that during a course of study the instructors taught only a wealth of non-related pieces of information. Even the mere storage of information in memory would be very difficult; a deeper understanding of what relationships (such as cause-and-effect relationships) are essential would be completely impossible. Also for the application of a theory, it is, of course, extremely important to know the corresponding *relationships* between concepts in order to use them for the explanation and prognosis of phenomena (see Sect. 2.4).

- A "good" theory provides *general knowledge* from which more specific insights for **concrete individual cases** can be derived. A "good" theory refers to an empirically validated theory. In the view of scientific realism (see Chap. 3), we assume that, after frequent validation of theories, there are good indications that their statements largely reflect the truth. Engineering and medicine provide exemplary applications of theoretical knowledge to practical problems, where general scientific knowledge is used to solve a specific problem (for example, selecting a material with particular properties, treating a malfunction of an organ). Corresponding applications in the field of marketing research are quite obvious. For example, if one knows the Elaboration Likelihood Model outlined above, then the next step to apply the theory to developing an advertising campaign is quite clear.
- Furthermore, theories can help derive suggestions and instructions for **further research**. First of all, we have to think about the empirical testing of the statements contained in a theory. In addition, the nature of the relationships shown in a theory (see above) is the subject of theoretical and empirical investigations. For example, what is the relationship between advertising budget and advertising impact (linear or non-linear)? In addition, measurements can also play an important role. How can we measure variables such as "relative product quality" or "success"?
- Ultimately, it goes beyond the practical usefulness, developing and testing theories serve the basic need of many (thinking) people to **understand** their surrounding reality. Thus, for centuries, people have been trying to understand how and why stars make their tracks in the sky. This happened long before any use for this knowledge (e.g., for space programs) was even thought of. Theories, with their capacity to organize knowledge, also evidently correspond to a human need for understanding the world.

In connection with the example of the ELM presented above, it has already been (implicitly) shown that there seems to be an *overlap* of "theory" and "model". Particularly in different areas of economics and business, researchers often work with models. What is the meaning of a "**model**"? In general, models are simplified representations of relevant parts of reality. For example, the influencing factors and their interaction, which are particularly relevant for communication effects, are parts of the ELM. There is an obvious simplification in the model, since the corresponding real psychical processes are related to a large number of other influencing factors, feedbacks, etc. and are much more complex. With the aid of such simplified representations one can describe, analyze, and design the essential elements of a (more complex) problem area.

In science and in everyday life, we deal with very different types of models, including:

- Graphical models (e.g., flowcharts, maps)
- Figurative models (for example a wooden model of a building for an architectural competition)
- Verbal models (for example, verbal descriptions of relationships)
- Mathematical models (e.g., regression model)

Figurative models are unlikely to play a role in marketing research, the other forms are very common. Thus, the ELM uses both graphical and verbal forms. Chapter 7 discusses aspects of modeling in more detail.

Demetris Portides (2008, p. 385) illustrates some other aspects of models:

"Despite the disparity of meanings in the use of the term 'model', we can discern that most, if not all, of its uses indicate that 'model' is strongly tied to representation, i.e. a model is meant to represent something else, whether an actual or an ideal state of affairs, whether a physical or an ideal system. For instance, a model of a building is a representation of an actual (or actualizable) building. Moreover, we can discern that 'model' is also strongly linked with idealization and abstraction, i.e. a model represents a physical system in an abstract and idealized way."

By discussing theories in this chapter, it has (hopefully) become clear that, at least from the perspective of scientific realism used in this book (see Chap. 3), theories are representations of real phenomena. In that sense, theories are a subset of models (in a general sense). Is this also the case the other way around; is every model a theory? The answer to this question is negative, because many models do not correspond to the features of theories formulated in this section. This becomes immediately clear when using the simple examples of a map or a wooden model (see above). There are also pure "measurement models" in marketing research that only serve to measure certain constructs (such as attitudes, motivation), but which no one would view as theory in the sense outlined here. Obviously, theories are a special form (a particular subset) of models, but many models do not fit the characteristics of a theory (Hunt 2010, p. 78).

2.2 Theory and Reality

Theories in marketing research have the function of facilitating the understanding of reality (although this does not become equally convincing for all theories). The term "**reality**" stands for real occurrences, for example, for the effects of an advertisement on the attitudes of customers, the actual consequences of a price increase, the innumerable purchase decision processes of consumers, etc.

Stathis Psillos (2007, p. 212) defines "reality" superbly briefly as:

"Reality: Everything there is.

The philosophical issue over reality concerns (1) its scope: what the elements (or constituents) of reality are and (2) its status: whether it exists independently of the human mind."

The philosophy of science position on so-called "scientific realism" (see Chap. 3) states that reality exists independently of the perspective and perception of the researcher (Godfrey-Smith 2003, pp. 173ff.). Jaccard and Jacoby (2010, p. 7) identify this position in the following way: "There is an external world comprised of objects that follow a myriad of natural facts and laws. It is up to us to discover these facts and laws. Using this perspective, science has evolved and prospected as an approach for gaining knowledge that mirrors the presumed actualities of the real world."

If one starts from the position that there exists a reality *independent* of the individual perception, and that this reality should be described and understood, then certain characteristics of the reality can be determined, which are important for appropriate—and not least empirical—research. Reality, according to Jaccard and Jacoby (2010, pp. 9–10) is:

- Complex,
- Dynamic,
- Mostly obscure and
- Unique.

A very simple example of a supermarket is used to explain these aspects:

Complexity The attempt to *fully describe* this supermarket will never succeed. Capturing all the details of this retail store (arrangement of shelves, lighting, inventory levels, locations of staff, walkways of customers, placement of products, etc.) at a particular time will overwhelm any and every extremely patient researcher, particularly the cognitive abilities of human beings.

Dynamics Even if it were possible to describe all the details of the supermarket, little would be gained, because the reality is *constantly changing*: new customers enter the store, shelves are refilled, it gets darker, etc. A full description of the reality at a certain time would be useless a short time later.

Concealment Numerous—even essential—aspects of reality are *not directly observable*. For example, for the situation in the supermarket, it is quite important to know what needs different customers have, how qualified and motivated the sales force is, how profitable the supermarket is, etc. All these aspects can only be determined with special measurement methods, but not through direct observation.

Uniqueness Because it is impossible to replicate a particular situation in the supermarket with identical clients that have stable needs and intentions, with identical shelf placements, identical external conditions etc., a comprehensive description or explanation would be useless, because *a particular situation does not repeat itself* such that one can use a formerly gained and very detailed knowledge.

Against this background it immediately becomes clear that it is hopeless to represent reality *completely* or almost completely through research and theory building. Rather, in (empirical) research, we look only at selected aspects of an overwhelmingly complex reality. Of course, this has important consequences for the research methodology, which is discussed in more detail in Chaps. 6–9.

Concepts, as explained in detail in the previous section, are important mental aids for a simplifying and, at the same time, abstracting consideration. With their use, we do not describe the full complexity of a particular situation, but focus on *relatively few abstract aspects* that are relevant for the problem of interest (largely independent of a specific situation). (See also Sect. 4.1.)

The use of concepts (and theories based on them) to simplify and abstract is illustrated here by referring to the above supermarket example:

Imagine that a (fairly simple) theory was built: "High attractiveness of store design" leads to "relatively large numbers of customers" and this in turn to "relatively large economic success". What happened here? We have abstracted from the many details of store design (size, floor plan, lighting, materials used, decoration, etc.) and instead we use only the single concept of "attractiveness", thereby simplifying this influencing factor of economic success by not considering further details. However, the problem arises of defining and measuring an abstract concept such as "attractiveness". Some of the following chapters will be devoted to this problem.

How does the use of concepts and theories affect the problems mentioned above?

Complexity: The typically strong complexity reduction has been explained above.

Dynamics: By abstracting from the details that change over time, one comes to statements that are largely independent of many changes over time.

Concealment: Since one no longer has to worry about a (too) wide variety of facets of reality, one can concentrate on measuring the central features (such as "store attractiveness"), applying sophisticated methods.

Uniqueness: By abstracting from the specifics of a given situation, one comes to more general knowledge that is useful in many similar (but not fully identical) situations.

In theories, concepts are systematically linked, that is, relationships are established between concepts (see Sect. 2.1). In this sense, a theory generalizes similar relationships between phenomena of reality that may be different in terms of various details, (for example, "attractiveness" → "demand"). This raises the question as to what extent the concepts used in a theory and their relationships to one another *correspond to reality*. Just as in a testimony in court that accurately reflects a process of action and events, or in a newspaper article in which a political event is reported accurately and impartially, a theory that accurately reflects the corresponding reality is said to be (at least approximately) **true**. Shelby Hunt (2010, p. 287) summarizes this plausible basic idea clearly and concisely:

"When confronted with any theory, ask the basic question: Is the theory true? Less succinctly, to what extent is the theory isomorphic with reality? Is the real world actually constructed as the theory suggests, or is it not?"

Section 5.3 presents, in greater detail, this criterion for the evaluation of a theory in the form of the "inductive-realistic model of theory tests". Particularly noteworthy is the question, "To what extent is the theory isomorphic with reality?". Shelby Hunt does not use the word "isomorphic" accidentally. The word "isomorphic" expresses the idea that the structure of a theory should correspond to reality, but should not reflect reality in full detail. This makes it clear that a complete match with reality is certainly not possible because every theory is and must be simplifying and abstracting. A theory that reproduces too many details of reality would be bound to these details and thereby (too) limited in its relevance. It also becomes apparent that in the perspective of scientific realism, we refer to "**approximate truth**", which Sect. 3.1 discusses in more detail.

The description of the truth of a theory according to Hunt (see above) clearly reflects the so-called **correspondence theory of truth** (for details see Haig and Borsboom 2012). This designation is easy to understand because Hunt's description refers to a "correspondence" of theory and reality. Thus, one can define "truth", but still has no indication to what extent theory and reality match. "The correspondence theory of truth, while well suited as a *definition* of truth, does not offer any *criteria* to evaluate the truth of a sentence" (Schurz 2014, p. 24). This is the task of empirical research, which is what this book is all about.

Brian Haig and Denny Borsboom (2012, p. 287) explain the relevance of "correspondence truth":

"We think that there are a number of reasons why correspondence truth matters in the quest to undertake and understand science. Correspondence truth provides scientists with an important orienting ideal for their research to approximate. Neither the fact that the ideal of truth cannot always be fully attained, nor the fact that we often do not have strong grounds for knowing how closely we have approximated the truth, counts against the importance of holding to correspondence truth. Ideals are maximally valued goals and are, therefore, important, even though they cannot be fully realized."

2.3 Lawlike Generalizations and Explanations

2.3.1 Lawlike Generalizations

As Sect. 2.1 shows, statements about lawlike generalizations are an essential part of theories. This aspect is deepened in this section. In the scientific explanations discussed in Sect. 2.3.2 below, theories and the statements they contain about lawlike generalizations are used in order to understand observations from reality.

The term "law" is often used with different meanings. On the one hand, a law refers to state regulations that are generally binding; non-compliance usually leads to sanctions. This aspect is less interesting in the context of this book. On the other hand, **laws** or **lawlike generalizations** describe (at least in the short term) stable relationships between certain phenomena in reality (including psychological and social phenomena) according to the pattern: (Almost) whenever "A" occurs, then "B" follows. In the scientific literature, a distinction is made between "lawlike generalizations" and "laws" (see, for example, Hunt 2010, pp. 143–144, see the quotation below); but this distinction is not relevant for the present introductory presentation. Therefore in this chapter we will mostly use the term "law". Laws are present in nature and will be *discovered* sooner or later (or maybe never). Most readers should be aware from their schooldays of "Ohm's Law", according to which the current strength in a conductor at constant temperature is proportional to the voltage. This law was discovered by Georg Simon Ohm (1789–1854). "Notice (. . .) that a law may currently be undiscovered (though I can't give you an example of one of those!) and that, after it has been discovered, it need not be officially called a 'law' (as with the axioms of quantum mechanics, Bernoulli's principle, and Maxwell's equations)." (Lange 2008, p. 203).

One speaks of a lawlike generalization / law when a certain regularity can be observed and substantiated or classified into a theoretical context. This is thus different from other relationships, both from logically compelling statements (for example: triangles have three sides, one full glass cannot be empty at the same time), and from coincidental relationships (for example, all waiters in a certain restaurant wear glasses). Laws rely on a certain need for coherence (Lange 2008, p. 204). Against this background, laws govern not only the respective events, but also corresponding processes at *other times* or in other situations.

Shelby Hunt (2010, pp. 143–144) characterizes lawlike generalizations and laws:

"In order for a generalized conditional statement to be a lawlike generalization or, alternatively, a lawlike statement, it must have (a) empirical content, (b) exhibit nomic necessity, and (c) be systematically integrated into a body of scientific knowledge. The empirical-content criterion successfully weeds out strictly analytic statements, tautologies, and nonsense generalizations from

(continued)

being considered lawlike. The nomic-necessity criterion serves the useful purpose of distinguishing lawlike statements from accidental generalizations such as 'All products with the trade name Maxwell House have a coffee base'. Finally, the systematic integration criterion enables us to differentiate lawlike statements from strictly empirical regularities. Empirical regularities have been shown to play an important role in the context of scientific discovery.

Lawlike generalizations become laws when a substantial body of corroborative empirical evidence has been developed."

In marketing research, we do not deal with deterministic contexts (as in Ohm's Law), which are always and unconditionally true, but rather with *probability statements* which do not allow for a case-specific certainty. Crucial for this is that for (empirical) marketing research, complex relationships of a variety of economic, behavioral, legal, etc. influencing factors and conditions apply, for which it is practically impossible to capture all the relevant influencing factors and their interactions. This is typical for many social science issues, but not just for that. For example, in the case of a natural phenomenon such as the formation of weather and corresponding prognoses, it shows that the great complexity of the interactions of different physical processes makes exact and reliable analysis and (especially long-term) prognoses more difficult. Because of this, one often finds corresponding probability statements (for example, the probability of rainfall in a region in a certain period of time). The same applies to medicine with its frequently occurring problems of an exact diagnosis and prognosis of successful therapy. In such cases one also speaks of the "lawlikeness of statistical generalizations" (Schurz 2014, p. 124), meaning that the lawlike relationship between traits (e.g., beer consumption and obesity) is not (in a deterministic way) in all corresponding cases observable, but only in a detectable proportion of cases. Section 2.3.2, which follows, shows some considerations on "statistical explanations".

Against the background of such complex interrelationships one often finds the reference to a **ceteris paribus clause** ("under otherwise equal conditions"), which serves to make it clearer in the analysis of the relationship of fewer variables, that keeping all other possible influencing factors as constant is assumed, for instance: "*ceteris paribus*, if demand for a given product exceeds supply, prices will rise." Here, it is obvious that the ceteris paribus clause is meant to ground the possibility of exceptions: "the law holds as long as all other factors (e.g., the existence of an alternative product) remain constant." (Psillos 2007, p. 38).

The above description of scientific laws and lawlike generalizations implicitly implies that these are relationships that, if given certain conditions, are *general*. The striving for *universal* correctness of statements (within certain limits) as opposed to the solution of concrete and thus more specific problems in practice is considered a *characteristic of science*.

A quote from Shelby Hunt (1976, p. 26) may illustrate the claim of science regarding the universality of statements:

"Every science presupposes the existence of underlying uniformities or regularities among the phenomena that comprise its subject matter. The discovery of these underlying uniformities yields empirical regularities, lawlike generalizations (propositions), and laws. (..) Uniformities and regularities are also a requisite for theory development since theories are systematically related sets of statements, including some lawlike generalizations, that are empirically testable."

What is the relevance of laws for marketing research?

- They are necessary for the *explanation* of phenomena. For example, if one knows the relationship between the commission pay share of sales force and sales efforts, then this knowledge can be used to explain changes in sales that have occurred (for the nature of scientific explanations, see Sect. 2.3.2).
- Furthermore, the knowledge of laws is a prerequisite for *predictions*. If one knows the relationship between two variables, then one can specify how a variable is likely to respond to the change in other variables. For example, an increase in sales volume is to be expected following a price reduction (for the relationship between explanations and predictions, see Sect. 2.4).
- Ultimately, the knowledge of appropriate laws is often a prerequisite for *influencing* success factors, which is the typical task of marketing management. If, for example, one knows the factors influencing the success of product innovations and their effects, these factors can then be shaped (see Sect. 2.4) to achieve the desired result. This includes the knowledge of the relevant lawlike generalizations and the will and the ability to influence the success factors. However, there are also numerous examples of managers who, without the knowledge of explicit laws or lawlike generalizations, make decisions with great success, for example, based on experience and intuition.

Given this relevance of laws, it is not surprising that their discovery and investigation alone form a considerable part of empirical research, even without focusing on more comprehensive theories.

The last part of this section characterizes more exactly scientific laws. According to Hunt (1991, 2010), there are four characteristic features of laws (see Fig. 2.2), whereby here, too, the focus is on marketing research. There may be other points of view in other disciplines. First of all, we have to deal with **general conditional relations**, that is, *if-then relations* in the sense of: "If A occurs, then B also appears". Considering the example of the Elaboration Likelihood Model (see Sect. 2.1), an example could be: "If the involvement is high and the ability to process information is given, then a comprehensive information processing (central route) occurs." Furthermore Hunt (2010, pp. 134ff.) demands **empirical evidence** of statements

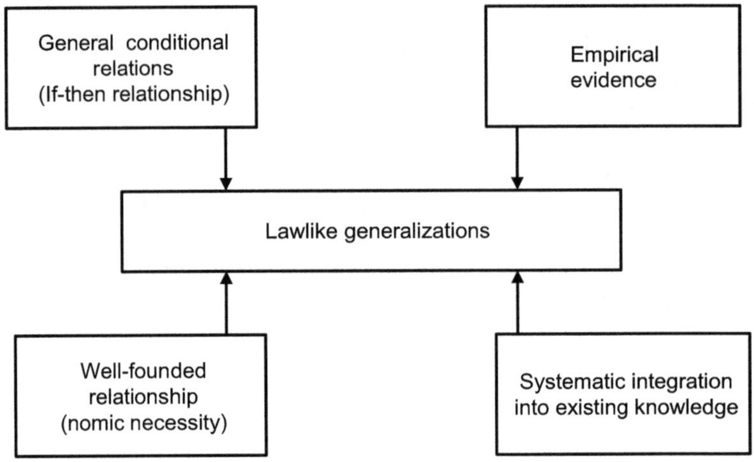

Fig. 2.2 Characteristics of lawlike generalizations according to Hunt (2010)

about laws. The extent of previous empirical tests with positive outcomes can be very different and will, of course, significantly affect confidence in explanations and predictions based on their respective laws. Reference is made here to the inductive-realistic model (Sect. 5.3) and to empirical generalizations (Sect. 9.3). The third characteristic of a law, according to Hunt (2010, pp. 136ff.), is that there must be a **well-founded relationship** ("nomic necessity"). This distinguishes a law from situational or random statements. Laws established in this way have significance beyond an individual case, which is of central importance for scientific statements.

Here is an example based on Psillos (2002, p. 8) for a relationship in which the first two characteristics mentioned above (if-then relationship, empirical evidence) are fulfilled, but are not well-founded (no nomic necessity):

At the moment of writing this passage, one of the authors of this book looked into his purse and found that all coins in it had a value of less than € 2, -. In this case, a conditional relationship applies: "If there is a coin in the purse of A.K. today, then it has a value < € 2, -." One can easily check the statement empirically, simply by looking into this purse. But would one speak of a law? Hardly, because obviously, this is a situationally determined random result.

The fourth requirement of Hunt (2010, pp. 138ff.) relates to the **systematic integration of statements**. Statements about laws should be—as far as possible—integrated into a larger system of statements (→ theory) in the sense that they are compatible with further relevant knowledge and, in this sense, without contradiction. However, this requirement hinders the discovery of entirely new and surprising findings that are not (yet) compatible with existing knowledge.

2.3.2 Scientific Explanations

The keyword "explanation" has already been mentioned several times. Its relevance is first illuminated with the help of two quotes:

> "Without explanation, there is no science" (Hunt 2010, p. 77).
>
> "The distinctive aim of the scientific enterprise is to provide systematic and responsibly supported explanations" (Nagel 1961, p. 15).

The meaning of explanations emphasized by these two authors is easy to understand: what would one think of astronomy if it were unable to explain a solar eclipse? What would one think of a botanist who cannot explain why no pineapples thrive in Alaska? What acceptance would marketing research have that could not offer any explanation for the impact of sales promotions?

> Ernest Nagel (1961, p. 4) identifies the meaning of explanations for science:
>
> "It is the desire for explanations which are at once systematic and controllable by factual evidence that generates science; and it is the organization and classification of knowledge on the basis of explanatory principles that is the distinctive goal of the sciences. More specifically, the sciences seek to discover and to formulate in general terms the conditions under which events of various sorts occur, the statements of such determining conditions being the explanations of corresponding happenings."

This does not mean that a scientific discipline can explain all relevant phenomena. If this was the case, further research would be superfluous. It also does not mean that all explanations are completely unequivocal and permanent. Rather, the history of science teaches that certain scientific knowledge from a previous point in time (for example, the idea that the Earth is the center of the universe) will be replaced by new and typically better insights. Nagel (1961, p. 15; see above) speaks of "systematic and responsibly supported explanations", *not of proven or certainly true* explanations. The fallibility of scientific knowledge is therefore accepted (→ "Fallibilism", see also Sect. 1.2).

The focus on explanations is not undisputed; the humanities are more focused on understanding. Schurz (2014, p. 14) characterizes such a position: "In the natural sciences, we *explain*, but in the humanities, we *understand*." This reflects the orientation of the natural sciences to the discovery and analysis of laws, while phenomena in humanities are not subject to (exact) laws. Understanding in this sense corresponds more to the complexity of the human mind. Of course, explanations can be an integral part of a broader understanding. Hunt (2002, p. 119) summarizes this view in one sentence: "Scientific understanding of a phenomenon implies, at least, that we can scientifically explain the phenomenon."

If one takes the position of scientific realism (see Chap. 3), then one might well agree with such a statement. However, this does not have to be the case for completely different approaches to the philosophy of science.

What is an **explanation**? It is the scientific approach to the answer to the question "Why ...?", the looking for reasons for a particular phenomenon. This already shows that explanations and causality (see Chap. 8) have a lot to do with each other. These relationships concern both the central idea ("What is the reason for ......?") and details of the corresponding conclusions. For an in-depth discussion of this aspect, please refer to Psillos (2002). Here are some examples of some "Why" questions:

- Why does a higher cumulative production volume lead to lower unit costs?
- Why do repeat purchases depend on customer satisfaction?
- Why is personal selling in the business-to-business sector so important?

For the explanation of such phenomena one makes use of the systematized knowledge in theories, including the respective laws. But there are certainly phenomena to be observed that cannot be explained that way because the available knowledge is not (yet) sufficient.

Scientific explanations consist of *three components*: at least one (deterministic or statistical) **law**, the given **conditions** in the given situation and the **phenomenon** that needs to be explained. For example, if there is a law that says that distractions of a target person who is exposed to an advertising message reinforce the message's effect—because the person's cognitive and critical processing of the message is restrained—then this law can be used together with a given condition (e.g., a person reading a message is distracted by music) to *explain* that in this case the message's effect was particularly strong. Therefore, explanations are *applications of (general) laws to specific cases*. The boundary conditions indicate whether the conditions for the effect of the respective law are given, and then a statement follows about the phenomenon to be explained. This implies that explanations (contrary to predictions) are typically linked to phenomena that have already occurred. Figure 2.3 illustrates these considerations. Here, we assume that the law used for the explanation is valid.

Scientific explanations in the social sciences must, according to Hunt (2010, pp. 78–79) and others, meet the following requirements:

- The phenomenon to be explained should *be expected* under the respective conditions. This relates to the principles of laws discussed above: if one wants to explain why phenomenon C3 has occurred under certain conditions (C1 and C2), then there must be a law or lawlike generalization which actually leads C1 and C2 to expect C3. If, for example, we want to explain why a bridge can withstand a maximum load of 50 tons, there must be statics laws that would allow such a load bearing capacity for the materials used, the existing subsoil, etc.
- The explanation must be *intersubjectively verifiable*, that is, it must be logically comprehensible and as far as possible free of prejudices, ideologies, etc. Usually, we assume that objectivity is a major goal in science. Hunt (2010, p. 77) cites a

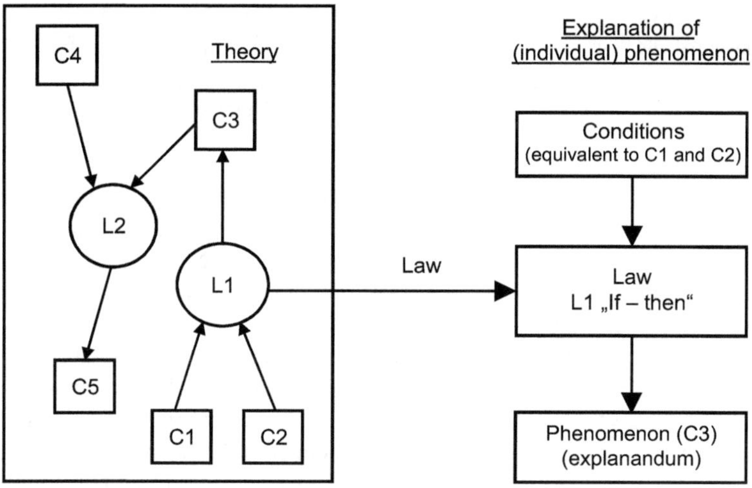

Fig. 2.3 Laws and explanations

statement by the philosopher Mario Bunge: "Science has not the monopoly of truth but only the monopoly of the means for checking truth and enhancing it."

- The explanation must find *empirical evidence*. This is to ensure that explanations refer to real phenomena.

Now to some typical forms of explanations. This presentation is based in parts on the corresponding overview in Hempel (1965, pp. 249ff; 1962). We use the following terms and symbols:

- **ES**: Explanans (set of conditions and laws that together explain the phenomenon in question)
- **ED**: Explanandum (phenomenon to be explained)
- **C**: Conditions (conditions that are the basis of the explanation)
- **L**: Laws, lawlike generalizations (universal, deterministic laws)
- **SL**: Statistical laws, lawlike generalizations (Probability statements, "If X .., then Y with q =. . .")

Deductive-Nomological Explanations (D-N Explanations)

Figure 2.4 shows the basic structure of deductive-nomological (derived from laws) explanations (D-N explanations). A classic example of such an explanation comes from Archimedes (285–212 BC): If the specific weight of a body is less than the specific weight of a liquid, then this body can float in the liquid (law). Thus, if the condition exists that the body's specific weight is less (or more), then one can explain why this body is swimming (or not swimming). This also relates to our expectations, for example, when we throw an object into water. This law applies to any case, that is, it is deterministic. If the explanans in a D-N explanation is given, the respective explanandum *has to* occur.

Fig. 2.4 Deductive nomological explanations

$$\left.\begin{array}{l} C_1, C_2, \ldots\ldots, C_k \\ L_1, L_2, \ldots\ldots, L_l \end{array}\right\} \text{ES (Explanans)}$$
$$\overline{\text{ED (Explanandum)}}$$

Explanations serve to assist in the understanding of processes and relationships. This requires not only the mere statement of a reason, but also the reference to laws and generalizations that constitute the relationship between the explanans and the explanandum (Psillos 2002, pp. 218–219). This indicates a way to test the truth of laws (or a theory). The D-N explanation refers to the statement that if certain conditions (B) are given and certain laws hold, that is, if the explanans (ES_i) has certain properties, the particular form of the explanandum (ED_i) occurs. But if this explanans ES_i is given and the corresponding explanandum ED_i does *not* occur, then obviously this is a contradiction to the assumed relationship. If no logical or methodological errors occur in such an investigation, then this explanation would be wrong. However, it will become clear that the claim that an empirical investigation should be free of methodological weaknesses poses considerable theoretical and practical problems for researchers (see Sect. 3.2 and Chap. 6).

Okasha (2002, p. 42) gives a simple example for an explanation:

"I am trying to explain why the plant on my desk has died. I might offer the following explanation. Owing to the poor light in my study, no sunlight has been reaching the plant; but sunlight is necessary for a plant to photosynthesize; and without photosynthesis a plant cannot make the carbohydrates it needs to survive, and so will die; therefore my plant died. This explanation fits Hempel's model exactly. It explains the death of the plant by deducing it from two true laws—that sunlight is necessary for photosynthesis, and that photosynthesis is necessary for survival—and one particular fact—that the plant was not getting any sunlight. Given the truth of the two laws and the particular fact, the death of this plant had to occur; that is why the former constitute a good explanation for the latter."

Deductive-nomological explanations are most likely to play a role in economics and business if they are based on scientific-technical laws. For example, in steel production, the share of energy costs per unit may be explained fairly clearly depending on the output volume. In the field of logistics, too, one can probably explain quite precisely how technical properties (such as speed or capacities) affect delivery times. For such statements we hardly need empirical research, because it can be better determined analytically. The domain of empirical research relates more to the effects and interactions of social (including economic) influencing factors (e.g., motivation → sales success, competence and trust → credibility). In these areas, however, there are hardly any deterministic relationships (such as "high market share *always* leads to high profitability"), because only in some exceptional cases are laws independent of situational factors and other boundary conditions. We might try to consider some of these conditions, but would then have to accept a great deal of complexity of the explanans, and yet there would still be uncertainty about the occurrence of the explanandum. We come back to this aspect when we discuss statistical explanations (see below) and refer to the remarks concerning a "ceteris paribus clause" in Sect. 2.3.

Against this background, in marketing research, a different kind of explanation is more important, the so-called **statistical explanations**. These contain at least one law of the form

$$P(A \mid B) = q$$

that is, the probability of A under the condition that B is given is equal to the value q. Statistical explanations are typical for the social sciences (including marketing research). We often have to deal with a very large number of influencing factors and complex interactions if, for example, we want to explain certain behaviors of customers. Because of this, we have to focus on relatively few influencing variables, which then *cannot explain* the explanandum *completely*, but only permit probability statements. "The idea (...) is that whenever we are confronted with a statistical explanation of a singular event, there are further facts such that, were they known, they could be used to afford a fully deductive explanation of the occurrence of the singular event in question." (Psillos 2002, p. 285).

More common in marketing research, however, is the interpretation of probabilities with respect to relative frequencies. Thus, for example, we can make relatively accurate statements regarding certain proportions relative to a larger number of cases (for example, "The proportion of ecologically conscious consumers is 11%."). "A natural objection might be that probabilistic-statistical laws can explain characteristics of large samples, but cannot explain anything about an individual case." (Psillos 2002, p. 244). For many practical applications, however, statements about proportions related to a larger number of cases do often suffice.

May Brodbeck (1968b, pp. 294–295) explains the possibilities and limits of statistical explanations:

"The social scientist (...) settles for something less than perfection. Completeness being far beyond his grasp, he renounces it as a goal. The renunciation has its price and its rewards. Which face will turn up when a die is cast is determined by numerous causes, the center of gravity of the die, the force with which it is thrown, and so on. An attempt to calculate the results of each throw by means of the laws of mechanics is practically hopeless, because of the difficulty in precisely measuring all the initial conditions. Instead we represent,

(continued)

as it were, the multiplicity of causes by a probability distribution for the attribute in question. The use of the statistical concept marks our ignorance of all the influencing factors, a failure in either completeness or closure or, usually, both. Similarly, the social scientist, deliberately selecting for a study fewer factors than actually influence the behavior in which he is interested, shifts his goal from predicting individual events or behaviors to predicting a random variable, that is, to predicting the frequency with which this kind of behavior occurs in a large group of individuals possessing the circumscribed number of factors. This is the price. The reward, of course, is that instead of helplessly gazing in dumb wonder at the infinite complexity of man and society, he has knowledge, imperfect rather than perfect, to be sure, but knowledge not to be scorned nonetheless, of a probability distribution rather than of individual events. After all, while we might much prefer to know the exact conditions under which cancer develops in a particular person, it is far from valueless to know the factors which are statistically correlated to the frequency of its occurrence."

Of the so-called "*statistical explanations*", we will shed some more light on the "inductive-statistical explanation" (I-S explanation) and the "explanation based on statistical relevance" (S-R explanation, see below).

Inductive-Statistical Explanations (I-S Explanations)

I-S explanations are similar to the representation in Fig. 2.5. The basic idea is that in a certain constellation of the explanans (certain conditions and corresponding statistical laws), a certain expression of the explanandum, is to be expected with *high probability*. Hence, it is an inductive reasoning. Also with I-S explanations there is obviously an expectation (→ "inductive") regarding the explanandum ED_i for a particular explanans ES_i. Marketing research rarely has corresponding examples. For example, we might think that a company has a certain probability of gaining a sales contract in competition with its competitors. When there is a large number of various contract opportunities in a period, then the probability of getting at least one of these contracts would be very large. However, such an explanation is certainly of very limited relevance.

What about the testing of I-S statements? Here the corresponding conclusions are a bit more complicated than for the D-N explanation (see above). If, for explanans ES_i, the explanandum ED_i actually occurred, the explanation is considered confirmed. But if there is a contradiction between an explanation ("If ES_i, then ED_i") and an observation ("ES_i is given, ED_i does not occur, but ED_j"), we need to engage in further considerations. The first step is to check the measurements with regard to

Fig. 2.5 Inductive-statistical explanation (I-S explanation)

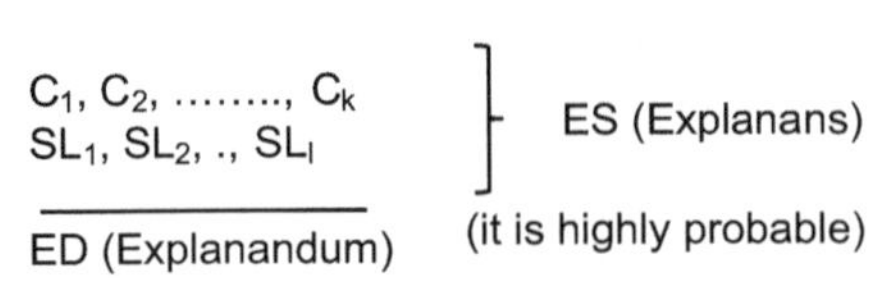

systematic errors (→ validity, see Chap. 6). If such a check does not provide any indication of such errors, it is still not clear why there is a contradiction between ES_i and ED_j: Either the explanation is wrong or the explanation is correct, but at random (it is "only" a statistical explanation) a contradictory result occurred. We say in such a case that "the explanation is not confirmed". I-S explanations are not falsifiable in the strict sense; the contradiction between explanans and the corresponding explanandum can be a random result. But there are "weak" falsifications, for example in the following way: "The probability that the explanation is wrong is very high." A corresponding conclusion is the basis of statistical tests (see Chap. 7), in which one decides on the acceptance or rejection of a hypothesis, if the probability of a random result—not justified by a systematic relationship—is small or large.

Explanations Based on Statistical Relevance (S-R Explanations)

In S-R explanations compared to I-S explanations, the requirements for the probability of the occurrence of a particular outcome are lower. One no longer assumes that this occurs with high probability as the result of a certain constellation in the explanans, but only that the corresponding conditions and lawlike generalizations have a recognizable or sufficiently clear impact on the explanandum. S-R explanations are therefore less informative than I-S explanations, but play a far greater role in empirical research. "There should be no doubt that the S-R model is a definite improvement over the I-S model" (Psillos 2002, p. 255). Figure 2.6 shows the formal representation of the S-R model.

Psillos (2002, p. 253) identifies the central idea of statistical relevance in the following way: An influencing factor B *contributes* to an explanation of the occurrence of an event A, if it holds

$$P\,(A|B) \neq P(A)$$

with P (A|B) = probability for A under the condition that B is given.

An example for S-R explanations relates to the purchase of organic ("green") products. We may find that the probability of buying such products is higher among environmentally conscious consumers than other consumers. Nevertheless, it is by no means the case that environmental consciousness (almost) exclusively determines the behavior of the aforementioned consumer group. Rather, environmental consciousness is only one factor influencing consumer behavior among others (for example, availability of financial resources, taste preferences). In this sense, environmental consciousness is relevant, but not necessarily the dominant or solely decisive factor. Its effect can most likely be shown in a larger number of cases (or in the form of appropriate probabilities); that is why we speak of *statistical* relevance.

Fig. 2.6 Explanations based on statistical evidence (S-R-explanations)

$C_1, C_2, \ldots, C_k$ } ES (Explanans)
$SL_1, SL_2, \ldots, SL_l$ }

Probability >> 0

ED (Explanandum)

Therefore the S-R explanations are actually no explanations in the sense that the occurrence of the explanandum (with high probability) would be expected. Hunt (2010, p. 91) refers to the example of the relationship between smoking and lung cancer. There are a great many smokers who do not die of lung cancer, and also non-smokers who die of lung cancer. Nonetheless, smoking appears to have an impact on the likelihood of developing lung cancer, so it is relevant in this sense, although (fortunately) not every smoker gets lung cancer. In this sense, S-R explanations are more about identifying more or less influential factors with respect to the explanandum. Such influencing factors can have different levels of importance for the respective explanandum and explain it more or less well. This refers to the idea of "**explained variance**" as discussed below.

In the empirical test of S-R explanations, the main question is whether the effect (measured in terms of a correlation, for example) of the independent variable on the phenomenon to be explained (dependent variable) is *sufficiently different* ("statistically significant", see Chap. 7) from zero. Again, the question arises of whether we deal with measurement errors in the empirical investigation. A (non-expected) low correlation could also be due to insufficient (erroneous) measurements. Furthermore, the substantial significance of tests should not be overestimated. For very large samples, almost every relationship—even a very weak one—between two variables becomes "significant", although its relevance may be very limited. "The claim that a significant correlation has been found between two features A and C is therefore a very weak claim, as long as we have no information about the sample size" (Schurz 2014, p. 196). Against this background, the significance of a relationship says little about its strength. For this, so-called "*effect sizes*" (see Sect. 7.2) are much more meaningful.

James Jaccard and Becker give an example of the difference between statistical significance and substantial significance (2002, p. 216):

"The poverty index in the United States for a family of four was defined in 1991 as an annual income of $ 14,120. Suppose that a researcher is interested in whether the mean 1991 income of a certain ethnic group differed from the official poverty level. The researcher examines this issue using data from a large national survey of 500,000 individuals from the ethnic group of interest. Suppose the observed sample mean for the ethnic group is $14,300.23. If the population standard deviation is also known, a one-sample z test can be applied. Suppose that application of this test leads to rejection of the null hypothesis. Then the researcher concludes that the mean income for the ethnic group population is 'statistically significant greater than the official poverty index'. Such a conclusion says nothing about how much greater the mean income is than the poverty index, nor does it say anything about the practical implications for the discrepancy."

S-R explanations are widely used in marketing research (and in the social sciences in general) because they typically deal with complex relationships between large numbers of variables. As a rule, we can only look at a small number of (especially) relevant variables and not fully grasp their complex interactions. For this reason, in empirical research concerning the explanation of a phenomenon (e.g., the growth of a market share or a change in preferences), we often focus on statements identifying the most important influencing factors, on the relevance and relative weights of the relevant influencing factors (e.g., in the form of regression coefficients; see Chap. 7), and on the types of relationships (e.g., positive / negative, linear / non-linear).

Explained Variance and Effect Sizes

In empirical research practice, one often looks at scientific explanations from a different perspective. Researchers use statistical methods that reveal the proportion of the variance of a dependent variable explained by one or more independent variables. Why do researchers consider "*explained variance*" when they want to explain, for example, profit, market share, growth, income etc.? The basic idea is quite simple: what is scientifically interesting is typically the difference in the values of these variables (e.g., profit) in different companies, people, etc. and one wonders why, for example, different companies differ in profit or innovativeness or different consumers save different parts of their income. The variance is just a measure for describing such differences. If it is possible to identify influencing factors (or independent variables) that influence the variations (differences) of the (dependent) variables of interest, then we evidently have identified reasons or explanations for these variations (differences). For example, if we find, based on an empirical analysis led by theoretical considerations, that the factors education, work experience, and duration of employment determine the variance of employees' incomes by 70%, then these *differences* in income are already *largely explained.* This example also shows the analogies of dependent variables and explanandum as well as of independent variables and explanans.

Measures such as the proportion of explained variance or any other effect sizes (see Sect. 7.2) are also interpreted as indicators for the quality of an explanation against this background. Some studies also use such effect sizes as measures of the state of knowledge in various fields (see, for example, Aguinis et al. 2011; Eisend 2015). If one can explain a large part of certain phenomena in a research field, then this field is probably more developed than other fields of research with lower levels of explained variance. Therefore, relationships with large effect sizes are also more informative for practical applications.

James Combs (2010, p. 11) briefly characterizes the relevance of effect sizes for science and practice:

"A theory might find support, but its explanatory power—that is, the effect size observed—is so weak that further efforts to develop the theory might not be warranted. Small effects also raise questions about managerial relevance. (...) If managers begin to act on theories that are supported by small effects, they are not likely to notice positive results even when they occur."

2.4 Explanations, Predictions and Strategy Development

Explanations, predictions and development of strategies are closely related. **Predictions** are—not surprisingly—about predicting future conditions and their implications. Here, the correspondence to the explanations discussed in Sect. 2.3 becomes immediately clear. The central idea was that after a certain constellation of conditions and laws (explanans), the appearance of a particular phenomenon (explanandum) follows. Thus, if the explanation is valid, one can conclude that the corresponding state is to be *expected* when this constellation occurs (→ forecast). Hence, any explanation would be a *potential* **prediction**. It is easy to see that explanations are, so to speak, "backward-looking", explaining previously observed phenomena just as they are, while predictions are (of course) oriented toward the future. In science and practice, a successful—that is, an accurate description of a future actual state—prediction is seen as a particularly strong evidence for an explanation. A famous example of this in history of science is Einstein's prediction of a deflection of light, which was confirmed during an eclipse in 1919 and became an impressive confirmation of his theory of relativity.

The aspect of **strategy** is more related to practical applications. It is about—on the basis of existing knowledge and experience—the decisions to influence future developments. Predictions develop expectations about the future; strategies try to *influence* these actively. In this case, the knowledge about relationships between the explanans and the explanandum is used in such a way that one knows how one must influence design conditions—assuming the validity of corresponding laws—in order to achieve a desired result. For example, knowing that in some markets early leadership leads to consistently superior profitability, it follows that a company must strive for early market leadership to achieve that goal.

The central ideas and differences are briefly described as follows:

- **Explanation**: Determination of the conditions and laws that have led to a specific fact.
- **Prediction**: Expectations for a future event based on known conditions and laws.
- **Strategy**: Manipulation of the conditions in order to achieve a desired state under given laws.

A very common application of predictions in empirical research is the *testing of hypotheses*. The basic idea here is that a hypothesis predicts what result—typically with regard to a relationship of variables—should occur if the theory from which the hypothesis comes is true. If the empirical result does not correspond to this prediction, the theory is not confirmed (assuming measurement errors or random errors are not the reasons for this result, see Sect. 6.3). This basic idea is discussed in more detail in Sect. 5.2 ("hypothetico-deductive method") and in relation to the inductive-realistic model of theory tests (Sect. 5.3).

Nevertheless, there are also objections to the equivalence of explanation and prediction. Thus, in practice, enough examples exist of successful predictions without appropriate explanations and, conversely, explanations without sufficient predictive power. For example, there are marketing managers who, based on their experience or intuition, can estimate fairly well (→ *prediction*) whether an applicant is a sales talent or not, but they could hardly explain exactly what constellation of characteristics led them to this prediction. On the other hand, the business press shows that the (subsequent) *explanation* of the successes and failures of companies is possible, but that a prediction of these developments is much less successful. Apparently there are differences between explanatory and predictive skills in (research) practice (see, for example, Psillos 2002, pp. 235–236).

Jaccard and Jacoby (2010, p. 16) give illustrative examples from everyday life to show that explanation and understanding in practice are not always connected:

"Although prediction and explanation often go hand in hand, the two are distinct. The person who tells the auto mechanic 'Every time I step on the car's accelerator, I hear a rattle in the engine—let me step on it and you can hear what I mean' may be able to predict without being able to explain. Similarly, as a moment's reflection about weather forecasting will reveal, being able to explain how the weather we experienced today came to be does not necessarily mean that we also are able to accurately predict the exact date and time when this precise weather will occur again."

For science and practice, it is worth noting whether a prediction refers to *individual cases* or to *proportions in groups*. In the first case, the basis for this would have to be a DN explanation (or an IS explanation, see Psillos 2002, p. 244), from which it can be (fairly) clearly deduced which state (explanandum) to expect for a particular explanans (see the above example of Archimedes' floating body). Such types of predictions hardly occur in marketing research. By contrast, predictions of proportions based on statistical explanations are more common, for example, "Under conditions x, y and z, at least 80% of our customers will most likely accept a 3% price increase".

May Brodbeck (1968a, p. 10) explains both types of predictions:

"It makes no difference whether the premises are statistical or deterministic, as nonstatistical generalizations are called. If they are deterministic, we may predict an individual event; if they are statistical, only statements about classes of events may be either explained or predicted."

Finally, a few remarks on strategy and related aspects of the relationship between theory and practice. As the beginning of this section showed clearly, strategy involves the use of knowledge about laws and theories for practical questions. However, the following factors typically limit such uses:

- Practical activities and decisions relate to very specific situations which are not fully represented by theoretical insights. The aspect of abstraction in theory formation explained in Sect. 2.1 leads to a certain degree of generalizability of the statements, but it must, so to speak, be "traded off" with limitations in regard to concrete situations.
- All predictions and forward-looking measures are, of course, associated with uncertainty about expected events or effects.
- Dynamics through human activity may lead to the fact that the laws based on past experience no longer apply to the same extent in the future. For example, one could imagine that the effect of market leadership on the cost position of a company decreases because several companies follow this strategy at the same time.

Against this background, the distinction in basic research and applied research, which is common not only in marketing research, becomes understandable. **Basic research** is not intended to solve directly practical issues, it deals rather with relatively general concepts and leads mainly to findings that serve the general understanding of the phenomenon of interest. In contrast, **applied research** focuses on a current (more or less) specific problem in a given situation, uses concepts that have a narrow focus, and leads to results that do not primarily lead to an increase in general knowledge (Jaccard and Jacoby 2010, p. 31).

Theory and practice are often considered as two different worlds. Practitioners often see theories as too abstract or too unrealistic to be helpful in solving practical problems. The different aims of theory (→ find the most general statements possible) and practice (→ solve special and concrete problems) seem to support this view. Furthermore, many theoreticians (here: academic marketing researchers) tend to pay less attention to some practical problems and focus on questions that are discussed in the respective academic community. It certainly plays a role that for success and a career in an academic field, the acceptance of the results of scientific work by reviewers, members of appointment committees, etc. is sometimes more important than the relevance of this work in regard to practical problems.

This issue has led to a discussion on **relevance** versus **rigor** (Varadarajan 2003). "Relevance" thus refers to the practical applicability of research results, which is sometimes operationalized in such a way that, in an investigation, on the one hand, variables that correspond to managerial decisions are used as independent variables, and on the other hand, it is important to what extent dependent variables are of interest to practitioners. The aspect "rigor", on the other hand, is oriented towards thorough theoretical foundation, careful methodological development and realization, sophisticated data analysis and adequate interpretation of the results (all according to the "state of the art" of research). Most of the time, this leads to a conflict between theoretical orientation with demanding (and thus often difficult to implement and understand) methodology (→ rigor) on the one hand and research with relatively concrete practical problems and robust methods (→ relevance) on the other.

Thomas and Tymon (1982) highlight—based on a comprehensive literature review and in relation to management research—five aspects that can contribute to a greater practical relevance of the research:

"**Descriptive relevance** refers to the accuracy of research findings in capturing phenomena encountered by the practitioner in his or her organizational setting" (p. 346).

"**Goal relevance** refers to the correspondence of outcome (or dependent) variables in a theory to the things the practitioner wishes to influence" (p. 347).

"**Operational validity** concerns the ability of the practitioner to implement action implications of a theory by manipulating its causal (or independent) variables" (p. 348).

"**Nonobviousness** refers to the degree to which a theory meets or exceeds the complexity of common sense theory already used by a practitioner" (p. 348).

"**Timeliness** concerns the requirement that a theory be available to practitioners in time to use it to deal with problems" (p. 349).

Regardless of considerations regarding the practical relevance of scientific research, the demand for the *empirical* confirmation of theories already mentioned in Sect. 1.2 (see also Chap. 5) establishes an essential connection to real processes and problems. Theories that are not adequately suited to explain and predict *real* phenomena do not do justice to the demands of a "good" theory and, if possible, should be replaced by a better theory.

Hunt (2002, p. 195) illustrates the relationship between theory and practice in the view of scientific realism through the juxtaposition of two formulations that, in his opinion, show "right" or "wrong" views:

Wrong: "It is all right in theory, but not in practice."

Right: "If it is not right in practice, it cannot be right in theory."

The perspective outlined here is also reflected in the well-known phrase, "Nothing is as practical as a good theory". This means that a "good" theory (see Sect. 5.1) can help solve a variety of different problems. For this, however, the focus on "good" theories is a prerequisite, that is, on theories where a reasonable fit with reality has already become clear in empirical studies.

2.5 Scientific Inferences: Induction, Deduction, and Abduction

Obviously, the development and critical examination of theories belongs to the core scientific tasks, not least because of the already outlined relevance of theories. Before explaining aspects of theory formation in Chap. 4, three approaches—in some cases fiercely discussed in philosophy of science—to the generation of scientific statements will be presented in more detail. The first of these are *induction* and *deduction*. Deduction also plays a central role in the empirical testing of theories and corresponding hypotheses (see Chap. 5). The end of this section outlines *abduction* as a conclusion that has gained more attention in recent decades.

Induction is the generalization of observed regularities in reality. For example, if we observe that an internationalization of marketing leads to higher profitability at a great variety of companies, then we might assume that there is a *general* relationship between internationalization and profitability and develop corresponding theoretical ideas. This is a conjecture from a number of special observations to more general statements or theories. To inferences in the opposite direction (from "general" to "particular") we will look later in this chapter and call them "deduction".

Chalmers (2013, pp. 42–43) formulates three minimum requirements for inductive reasoning, which also reveal some of the associated problems:

- The *number of observations* used as a basis for generalization must be large. But what does "large" mean here? How many observations would be necessary—10, 100 or 1000? In empirical research we try to solve this problem by applying sampling theory. This allows statements about to what extent observations are representative or with what probability relationships between variables are systematic or random relationships.
- The observations need to repeat under *different conditions* and lead to similar results. Given a general relationship, for example, between attitudes and behaviors, then this relationship must hold under a variety of conditions: under time pressure, for elections or purchase decisions, for low or high interests in the respective decision, etc. How many and which conditions would be necessary to come to a general statement? In empirical research, this problem is dealt with under the keywords "external validity" (see Sect. 8.3.2) and "generalizability" (see Chap. 9).
- *None of the observations should contradict* the derived general law. Since one hardly has to deal with deterministic contexts in the social sciences (including marketing research), somewhat weaker requirements apply here. But at least the number of contradictory observations must be so small (how small?) that the

probability for the validity of the derived statement is high (how high?). In empirical research, one uses appropriate statistical tests to decide whether the available data (number of cases, effect sizes) confirm a statement with sufficient confidence (see Chap. 7).

Okasha (2002, p. 20) provides an example of the widespread use of inductive modes of conclusions that applies even in everyday life:

"When you turn the steering wheel of your car anticlockwise, you assume the car will go to the left not the right. Whenever you drive in traffic, you effectively stake your life on this assumption. But what makes you so sure that it's true? If someone asked you to justify your conviction, what would you say? Unless you are a mechanic, you would probably reply: 'every time I've turned a steering wheel anticlockwise in the past, the car has gone to the left. Therefore, the same will happen when I turn the steering wheel anticlockwise this time.' (...) This is an inductive inference, not a deductive one. Reasoning inductively seems to be an indispensable part of everyday life."

Inductive reasoning is characterized by the way that we infer from existing observations to future—not yet observed—phenomena and, starting from a limited number of observations, end up with generalizations. For example, if in the past a correlation between two variables has been shown many times, then one can expect that it will also occur in a future observation; but if this relationship actually reappeared, then just this last observation is already in the past and a safe statement about further (future) corresponding observations can only be made if it is ensured that the future and the past will be the same, which, of course, is impossible (Schurz 2014, p. 50).

Gerhard Schurz (2014, p. 50) briefly summarizes the above-mentioned problem:

"Inductive generalizations are fundamentally uncertain (...). This is because the premises of an inductive generalization tell us only about the cases observed so far, while the conclusion generalizes to all and, in particular, to all future cases. For this reason it is also said that inductive conclusions are content expanding. Only in circumstances or worlds which are sufficiently uniform, whose future and past are sufficiently uniform, can we reliably infer the truth of an inductive conclusion from the truth of the premises. Nothing can logically guarantee that the future course of events will be sufficiently similar to the course of events observed so far."

David Hume (1711–1776) had already formulated this logical problem of inductive reasoning with respect to the claim of *certainty* of the statements (Newton-Smith 2000). In the second half of the twentieth century, Karl Popper's critique of inductive reasoning became particularly prominent. Popper (2002, pp. 3ff.) assumes, too, that there is no logical and compelling way to ascertain the truth of theories by inductive means. As a result, his considerations boil down to the fact that scientific theories always retain the character of conjectures, even after many observations that are consistent with them, as long as the theory is falsified by contradictory empirical results. However, it should be noted that this is about obtaining statements that are *certainly* true. The approach of scientific realism (see Chap. 3) will show that in its view, inductive conclusions are accepted, but at the cost of some uncertainty and inaccuracy ("approximate truth") of the statements. This is also the basis of the inductive-realistic model of theory testing, which plays an essential role in Chap. 5.

Karl Popper (2002, pp. 3–4) explains his rejection of an inductive reasoning as follows:

"It is usual to call an inference 'inductive' if it passes from singular statements (sometimes also called 'particular' statements), such as accounts of the results of observations or experiments, to universal statements, such as hypotheses or theories.

Now it is far from obvious, from a logical point of view, that we are justified in inferring universal statements from singular ones, no matter how numerous; for any conclusion drawn in this way may always turn out to be false: no matter how many instances of white swans we may have observed, this does not justify the conclusion that all swans are white.

The question whether inductive inferences are justified, or under what conditions, is known as the problem of induction."

In addition to the outlined logical problem, there are also more practical limitations to the formation of theory by induction, which Sankey (2008, pp. 249–250) points out:

First of all, the question arises as to whether the observations from which it is intended to generalize have really originated independently of an already existing theory. It is rather typical that certain preliminary information is necessary for the collection of such observations, that is, that these observations are not made at random but with a goal in mind (with regard to an emerging theory?). "A background of knowledge, which may include theoretical knowledge, must already be in place before the work of data collection may even begin." (Sankey 2008, p. 250) This issue will be discussed later in this book under the heading "Theory-ladenness" (see Sect. 3.2).

A *second* problem is that statements of theories often do not refer to (directly) observable phenomena; in marketing research, for example, they may refer to consumer attitudes towards advertising or the innovation orientation of companies. In that sense, corresponding parts of theories probably do not arise only through a generalization of observations.

A rejection of inductive reasoning would also lead to far-reaching limitations in the practical application of scientific knowledge, for example, in medicine, engineering and not least in the empirically oriented parts of marketing research. If one cannot infer from a limited number of previous observations (e.g., after the trial of a new medical therapy) the future validity of corresponding results (inductive), then one cannot say anything about the expected effects of a medication, the future performance of a machine, or the future effect of a price reduction. But there is hardly any claim in marketing research to make absolutely precise and certain statements; one tends more to an "epistemic modesty" (Schurz 2014). At least the empirical methodology (see Chaps. 6–9) sets some limits. Typically existing measurement errors and the nature of inferential statistics lead to inaccuracies and uncertainties of results. In Sect. 4.3.4 an empirical technique will be presented that is frequently used to generalize on an inductive way from a limited number of observations to more general statements, the so called "empirical generalizations".

Deduction is, in a sense, the counterpart to induction. One does not infer universally laws from a multitude of individual cases, but with the help of logical rules one derives from *general* statements other statements that refer to *more specific* cases. If, for example, there is a *general* positive relationship between motivation and performance, then one could derive (more sophisticated: "deduce") from this that in a more specific case, a positive relationship between the motivation and the performance of sales representatives should exist. In the case of a deductive conclusion, it is therefore clear—applying the logical rules—that given the appropriate conditions, the conclusion is also true; the deductive conclusion is thus *truth-preserving*. But that also means that this way of conclusion drawing does *not create new knowledge*. The result of the inference is already implicit in the given premises (Psillos 2007, p. 58).

This deductive reasoning can be used not only for the development, but also for the *testing* of theories. A common way to test theories is to derive corresponding statements (hypotheses, see Sect. 5.2) the correctness of which may be checked by the results expected / predicted on this theoretical basis confronted with actual observations. If there is a high degree of agreement we speak of an acceptance, otherwise of rejection (falsification) of the respective hypothesis and we question the theory on which the hypothesis is based. Section 5.2 discusses this so-called "**hypothetico-deductive**" approach in more detail. The (different) application of deduction in theory formation will follow in Sect. 4.2. In this case, it is about deriving a more specific theory from a more general one.

Now to the *comparison* of induction and deduction. The above description suggested that the two methods of scientific conclusion play different roles in theory formation and testing. Characteristic for induction is indeed the conclusion from a large number of individual cases to general statements (laws and theories); Schurz (2014, pp. 50–51) uses the term "**inductive ascent**". In the deduction it is the other way round: More specific statements are derived from general statements, which Schurz marks with the term "**deductive descent**". Figure 2.7 summarizes these two aspects. In addition, both types of conclusions differ with regard to their impact on knowledge development. Induction creates new (but not certain) knowledge;

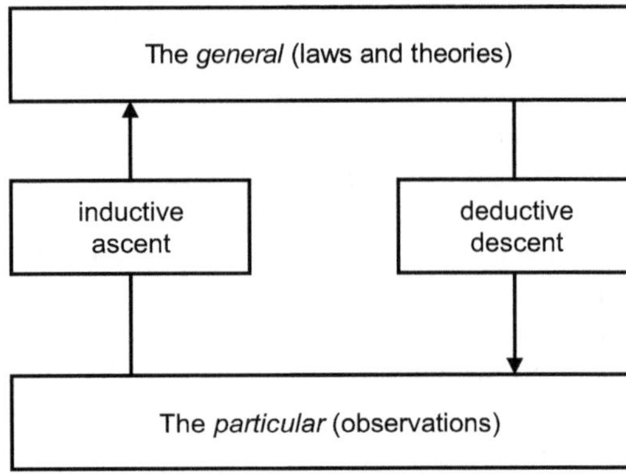

Fig. 2.7 Inductive-deductive schema (Schurz 2014, p. 51)

deduction, on the other hand, transfers already existing general knowledge (with certainty) to corresponding more specific cases. In this sense, induction is **knowledge-expanding** and deduction is **truth-preserving**.

Likewise knowledge-expanding, but of a different kind than induction, is **abduction**. This is about the conclusion of observations on their suspected reasons. Abduction is a "mode of reasoning which produces hypotheses such that, if true, they would explain certain phenomena" (Psillos 2007, p. 4). For example, one concludes in everyday life from the observation of a skid mark on a road and a demolished guardrail that an accident has happened. Although the observations could have other reasons (for example, it would logically also be possible that the skid mark originated from a full braking without accident and a truck could independently have damaged the guardrail in a turning maneuver), one decides on the presumed cause for the most plausible (or "best") explanation. Of course, this decision for a specific explanation is quite preliminary and needs further testing. A characteristic scheme of abduction goes in the following way (see Schurz 2014, p. 56):

1. A fact F (e.g., an event or a state) should be explained.
2. There is a certain background knowledge W that makes a certain explanation E plausible for the fact F.
3. Abductive conjecture: E is true.

Brian Haig (2009, p. 220) gives a short characterization of abduction:

"I take abduction to involve reasoning from puzzling facts to theories that might explain them. As such, abduction is a process of hypothesis or theory generation that can, at the same time, involve an evaluation of the initial plausibility of the hypotheses and theories proposed."

The abductive conjecture is a plausible (but not yet tested) hypothesis (see Sect. 5.2). Psillos (2007, p. 181) defines "**plausibility**" as "a feature of a hypothesis on the basis of which the hypothesis is deemed intuitively acceptable before any empirical evidence for it is being sought". In this sense, such a hypothesis would correspond to the best explanation (at a particular time). In the literature, therefore, abduction is also linked to "*inference to the best explanation*" (see, for example, Douven 2011; Lipton 2008).

Here is a simple marketing example for abduction:

Fact: A company suffers a continuous decline in sales in one year.

Background knowledge: It is well-known that a large number of existing customers place great value on high quality. However, the company had problems in securing product quality due to frequent personnel changes and unreliability of suppliers in the time in question.

Abductive conjecture: The quality problems are the cause of the decline in sales.

It is obvious that abductively drawn inferences cannot lead to *certain* statements. As in the case of induction, one assumes some degree of justification for these conclusions, but of course does not raise the claim of well confirmed truth. This would require further theoretical foundation and successful empirical testing. In this sense, induction and abduction produce new but not (yet) certain knowledge. In the case of deduction it is the other way round: the results of such inferences are indeed certain, because they are the logical derivation from given premises, but in the end they do not really go beyond the previous state of knowledge.

Abductive inference has become a matter of course for many people in everyday life, in professional practice and also in science. For applications in science Schurz (2014, pp. 55–56) refers to the example of Newton, who concluded—from the movement of planets around the sun—by abduction, the existence of a gravitational force. This example also highlights the role of abduction in theory formation. In this sense, the focus of abduction is the development of new ideas and hypotheses.

References

Aguinis, H., Dalton, D. R., Bosco, F. A., Pierce, C. A., & Dalton, C. M. (2011). Meta-analytic choices and judgment calls: Implications for theory building and testing, obtained effect sizes, and scholarly impact. *Journal of Management, 37*(1), 5–38.

Bagozzi, R. P. (1980). *Causal models in marketing*. New York: Wiley.

Brodbeck, M. (1968a). General introduction. In M. Brodbeck (Ed.), *Readings in the philosophy of the social sciences* (pp. 1–11). New York: Macmillan.

Brodbeck, M. (1968b). Methodological Individualism—Definition and reduction. In M. Brodbeck (Ed.), *Readings in the philosophy of the social sciences* (pp. 280–303). New York: Macmillan.

Chalmers, A. (2013). *What is this thing called science?* (4th ed.). Indianapolis, IN: Hackett.

Combs, J. G. (2010). Big samples and small effects: Let's not trade relevance and rigor for power. *Academy of Management Journal, 53*(1), 9–13.

Douven, I. (2011). Abduction. In E. Zalta (Ed.), *The Stanford encyclopedia of philosophy*. http://plato.stanford.edu

Eisend, M. (2015). Have we progressed marketing knowledge? A meta-meta-analysis of effect sizes in marketing research. *Journal of Marketing, 79*(3), 23–40.

Godfrey-Smith, P. (2003). *Theory and reality—An introduction to the philosophy of science*. Chicago: University of Chicago Press.

Haig, B. (2009). Inference to the best explanation: A neglected approach to theory appraisal in psychology. *American Journal of Psychology, 122*(2), 219–234.

Haig, B., & Borsboom, D. (2012). Truth, science, and psychology. *Theory & Psychology, 22*(3), 272–289.

Hempel, C. G. (1962). Two models of scientific explanation. In R. Colodny (Ed.), *Frontiers of science and philosophy* (pp. 9–19). Pittsburgh, PA: University of Pittsburgh Press.

Hempel, C. G. (1965). *Aspects of scientific explanation*. New York: Free Press.

Hunt, S. D. (1976). The nature and scope of marketing. *Journal of Marketing, 40*(July), 17–28.

Hunt, S. D. (1991). *Modern marketing theory: Critical issues in the philosophy of marketing science*. Cincinnati, OH: South-Western.

Hunt, S. D. (2002). *Foundations of marketing theory. Toward a general theory of marketing*. Armonk, NY: M. E. Sharpe.

Hunt, S. D. (2010). *Marketing theory—Foundations, controversy, strategy, resource-advantage theory*. Armonk, NY: Sharpe.

Hunt, S. D. (2015). Explicating the inductive realist model of theory generation. *Academy of Marketing Science Review, 5*, 20–27.

Jaccard, J., & Becker, M. (2002). *Statistics for the behavioral sciences* (4th ed.). Belmont, CA: Wadsworth.

Jaccard, J., & Jacoby, J. (2010). *Theory construction and model-building skills – A practical guide for social scientists*. New York: Guilford.

Kerlinger, F., & Lee, H. (2000). *Foundations of behavioral research* (4th ed.). Melbourne: Wadsworth.

Kotler, P. T., & Keller, K. L. (2012). *Marketing management* (14th ed.). Boston: Pearson.

Lange, M. (2008). Laws of nature. In S. Psillos & M. Curd (Eds.), *The Routledge companion to philosophy of science* (pp. 203–212). London: Routledge.

Lipton, P. (2008). Inference to the best explanation. In S. Psillos & M. Curd (Eds.), *The Routledge companion to philosophy of science* (pp. 193–202). London: Routledge.

Nagel, E. (1961). *The structure of science. Problems in the logic of scientific explanation*. New York: Harcourt Brace Jovanovich.

Newton-Smith, W. (2000). Hume. In W. Newton-Smith (Ed.), *A companion to the philosophy of science* (pp. 165–168). Oxford: Blackwell.

Okasha, S. (2002). *Philosophy of science—A very short introduction*. Oxford: Oxford University Press.

Petty, R. E., Cacioppo, J. T., & Schumann, D. W. (1983). Central and peripheral routes to advertising effectiveness: The moderating role of involvement. *Journal of Consumer Research, 10*(September), 135–146.

Popper, K. (2002). *The logic of scientific discovery*. New York: Routledge.

Portides, D. (2008). Models. In S. Psillos & M. Curd (Eds.), *The Routledge companion to philosophy of science* (pp. 385–395). London: Routledge.

Psillos, S. (2002). *Causation & explanation*. Durham: Acumen.

Psillos, S. (2007). *Philosophy of science A–Z*. Edinburgh: Edinburgh University Press.

Rudner, R. (1966). *Philosophy of social science*. Englewood Cliffs, NJ: Prentice-Hall.

Sankey, H. (2008). Scientific method. In S. Psillos & M. Curd (Eds.), *The Routledge companion to philosophy of science* (pp. 248–258). London: Routledge.

Schurz, G. (2014). *Philosophy of science. A unified approach*. New York: Routledge.

Thomas, K., & Tymon, W. (1982). Necessary properties of relevant research: Lessons from recent criticisms of the organizational sciences. *Academy of Management Review, 7*, 345–352.
Varadarajan, R. P. (2003). Musing on relevance and rigor of scholarly research in marketing. *Journal of the Academy of Marketing Science, 31*(4), 368–376.

Further Reading

Bird, A. (1998). *Philosophy of science*. Montreal: McGill-Queen's University Press.
Hunt, S. D. (2010). *Marketing theory—Foundations, controversy, strategy, resource-advantage theory*. Armonk, NY: Sharpe.
Psillos, S. (2002). *Causation & explanation*. Durham: Acumen.

The Philosophy of Science Viewpoint: Scientific Realism

3

3.1 Characterization of Scientific Realism

After discussing the essence and relevance of theories as a central area of science in the preceding chapter, the following chapters deal with the question of how theories can be developed and how the value of theories can be examined and assessed. To this aim, empirical research is of major importance. The demands on theories, and thus the standards for the assessment of their quality as well as for the assessment of their meaningfulness, are the subject of *philosophy of science* considerations. "The discipline of philosophy of science investigates how scientific knowledge works—its goals and its methods, its achievements and its limitations" (Schurz 2014, p. 1). Against this background, it is useful for the discussions that follow, to explain the philosophy of science viewpoint underlying this book: **scientific realism**.

Among others, the following authors emphasize that scientific realism has a strong or even dominant role in philosophy of science, respectively in marketing research:

Putnam (1975, p. 69): "The positive argument for realism is that it is the only philosophy that doesn't make the success of science a miracle."

Leplin (1984, p. 1): "Scientific realism is a majority position whose advocates are so divided as to appear a minority."

Hunt (1990, p. 13): "Many marketing researchers, either explicitly or implicitly, already are guided by scientific realism."

Hunt and Hansen (2010, p. 124): "(. . .) scientific realism seems to make the most sense for marketing, for no other philosophy is coherent (without being dogmatic), is critical (without being nihilistic), is open (without being anarchistic), is tolerant (without being relativistic), is fallible (without being

(continued)

M. Eisend, A. Kuss, *Research Methodology in Marketing*,
https://doi.org/10.1007/978-3-030-10794-9_3

subjectivistic) and—at the same time—can account for the success of science. It is a good candidate for providing a philosophical foundation for marketing research."

Haig (2013, p. 8): "Although the subject of considerable debate, and opposed by many antirealist positions, scientific realism is the dominant philosophy of science today. It is also the tacit philosophy of most working scientists. This fact, combined with its current heavy emphasis on the nature of scientific practice, makes scientific realism a philosophy for science—not just a philosophy of science."

The approach of scientific realism (abbreviated as **SR** in this chapter) has developed in the philosophical literature since the 1960s. This approach follows as a reaction to two other positions that played a significant role in the second half of the twentieth century: critical rationalism and relativism. These two positions will not be discussed here in detail, but we recommend that the reader looks at the relevant literature (for example Brown 2012; Chalmers 2013; Godfrey-Smith 2003; Schurz 2014), where reference is also made to the original sources. However, it is useful for the understanding of SR to know some central ideas of critical rationalism and relativism because some aspects of SR become particularly clear when *contrasting* them with the other approaches. For this reason, a few highlights of both critical rationalism and relativism are outlined here. Much simplified, some of the essential aspects of SR appear as a continuation or modification of ideas of critical rationalism, though it is largely opposed to relativism.

Firstly, we will look at **critical rationalism**. Karl Popper (1902–1994), who strongly influenced the philosophy of science over many years, founded and decisively shaped critical rationalism. In the context of this chapter, the following three aspects are important:

- Evaluation of knowledge by *falsification* attempts and not by inductive conclusions.

For centuries, so-called **induction** (for some details see Sect. 2.5) has played a major role in various scientific disciplines. Induction involves the generalization of regularities observed in reality. Induction includes two steps (Sankey 2008, p. 249):

- The collection of empirical data related to a phenomenon of interest
- The formulation of laws, rules and, in the next step, theories based on the generalization of observed regularities and relationships (see Chap. 4).

Thus, researchers concluded from a multitude of observations/studies that showed a certain effect that this effect exists *in general.* An example of such an inductive approach is the PIMS (Profit Impact of Market Strategies) study, also influential in marketing research, in which so-called success factors were identified

(Buzzell and Gale 1987) based on a large number of analyzed strategic business units.

There is broad agreement, irrespective of various positions in the philosophy of science, that inductive methods do not lead to *absolutely certain* insights (see Sect. 2.5). Nevertheless, an inductive approach in both everyday life and science is commonly applied and accepted, albeit without the full certainty of knowledge (see, for example, Schurz 2014, pp. 54ff. and Okasha 2002, pp. 19ff.). Popper (2002b; first edition 1935) rejected the possibility of obtaining insights by induction, and promoted empirical research that aims for **falsification** attempts as a way of examining scientific theories. Existing theories are continually exposed ("as strictly as possible") to falsification tests and are, depending on the result, either *temporarily* maintained or rejected.

One of the problems of Popper's falsification approach is that it is rarely practiced: falsified findings are not paid sufficient attention to. Researchers are typically driven by their personal motivation and the requirements of a scientific career not to falsify hypotheses and theories, but rather to find new theoretical solutions and to confirm them by appropriate empirical research. Furthermore, the applicability of the falsification approach is limited by the fact that the results of empirical studies with the aim of testing of theories may be flawed (see Sect. 3.2). If an empirical result does not agree with the corresponding theoretical conjectures, one does not know with sufficient certainty whether this is due to measurement errors or due to an incorrect theory (see, for example, Psillos 2007, pp. 71–72).

- *Provisional character* of scientific knowledge

The previous paragraph suggests that, in the view of critical rationalism, the ongoing critical (!) questioning of previously accepted theories and the development of better theories are central tasks of scientific research. Instead of a past position that suggests that science should gain secure and simultaneously true findings ("Fundamentalism", Phillips and Burbules 2000, pp. 5ff.; Schurz 2014, p. 3), with critical rationalism a **fallibilistic** approach has taken place, "which concedes that our understanding of reality is basically *fallible*, and our scientific knowledge can be more or less well confirmed, but it cannot be guaranteed to be free of error" (Schurz 2014, p. 3). This fallibilistic view is also part of SR (see further below).

- The position of *realism*

By "**realism**", philosophy of science refers to a position characterized by the assumption that reality exists, *independent* of the perception and interpretation of the particular observer (for example, Schurz 2014; Devitt 2008; Psillos 2006). Popper (2002a, b) proposes this view, sometimes implicitly, sometimes explicitly. This position may seem obvious and is present in SR, but the following remarks related to relativism (see below) show that in the literature divergent views have existed and still exist.

Karl Popper (2002a, p. 157) on theories and reality:

"Theories are our own inventions, our own ideas; they are not forced upon us, but are our self-made instruments of thought; (...) But some of these theories of ours can clash with reality; and when they do, we know that there is a reality; that there is something to remind us of the fact that our ideas may be mistaken. And this is why the realist is right."

Following these remarks on critical rationalism, we will move on to another philosophy of science viewpoint, which played a significant role in marketing research in the last third of the twentieth century and is still occasionally present today: "**relativism**". This term summarizes a spectrum in terms of justification and consequences of quite different positions (for an overview, see Swoyer 2003), all of which have one central idea in common: "Epistemic relativism is the view that claims to knowledge are invariably bound by particular historical, cultural frameworks and are true or legitimate only relative to their conditions of production." (Baghramian 2008, p. 236). It becomes immediately clear that this view negates the possibility of *objective* knowledge about reality. The two aspects that are particularly significant here are briefly explained below:

- *Context dependency* of findings

The key phrase "context dependency" that characterizes relativism refers to the influence of social, political, economic, religious, etc. factors that influence scientific research and theorizing. Thomas Kuhn (1922–1996) particularly notes this aspect in his analysis (1970, first edition 1962). Based on historical examples, Kuhn dealt with the influence of contextual conditions (social, political, intellectual, etc.) on the research process and coined the now famous (and almost popular) term **paradigm** as the "entire constellation of beliefs, values, techniques, and so on shared by the members of a given community" (Kuhn 1970, p. 175). For details on this important aspect, we recommend a look into further readings (for example, Carrier 2012; Hunt 2003; Psillos and Curd 2008; Rorty 2000) and, of course, the work of Thomas Kuhn himself (1970 and the 2012 new edition).

A *paradigm shift* can lead to a new interpretation of observations and thus to a new theory. A well-known example, from the history of science, for transition from one paradigm to another is the change from the Ptolemaic worldview (Earth as the center of the universe) to the Copernican worldview (Earth revolving around the sun). This change in the worldview led to completely new theories about orbits of planets, etc. As we now know, the "old" worldview was heavily influenced by the social context, and in particular by religious views. In economics too, the orientation or non-orientation along the paradigm associated with the concept of "homo oeconomicus" leads to fundamentally different theories and research methods. This also suggests that different social positions and interests influence the degree of acceptance of certain theories. In this view, scientific knowledge would be *context-dependent* and the claim of a systematic approach to an "objective truth" would be wrong. Relativists typically assume that such context dependencies largely determine scientific knowledge. Even if one does not take a relativistic position, one can hardly deny that the context of science (for example, political or ideological frameworks, or the power of sponsors of science) can influence, to some extent, scientific topics and results. We will come back to this point when we discuss the inductive-realistic model (see Sect. 5.3).

- Relation to *reality*

In many, but not all, cases relativists question or deny the existence of a reality independent of the observer (see above) (Godfrey-Smith 2003, pp. 181ff.). There are two aspects to be distinguished:

- The ability of researchers to observe and interpret real phenomena and processes is often influenced by certain assumptions, previously accepted theories or paradigms (see above). One can hardly imagine that someone makes scientific observations and collects data without any prior knowledge and without the influence of his or her mental and social environment. At least a selection of the studied phenomena of reality is required for a study. This problem of a biased perception and interpretation of reality, which limits the correctness of observations, is independent of the various positions within the philosophy of science and will be revisited in Sect. 3.2.
- Much more influential is the (constructivist) notion that not only is the *perception* and interpretation of reality subject to the influence of existing paradigms and theories, but that the processes of scientific discussion and theory development influence mental and social *reality* itself. Kuhn (1970, p. 121) points to this view: "Though the world does not change with a change of paradigm, the scientist afterward works in a different world". This approach finds its extreme expression in so-called radical **constructivism**, which is hardly represented today in marketing research. This view follows the basic idea that the human perception of reality is not passive, but rather is the result ("a construction") of cognitive processes. Richard Boyd (1984, p. 43) summarizes this position—which he does not represent—in one sentence: "Scientific methodology is so theory-dependent that it is, at best, a construction procedure, not a discovery procedure". From this it is falsely concluded that reality itself does not have a "definite and mind-independent structure" (Psillos 2006, p. 688), but that reality is constructed (for logical problems of radical constructivism see Schurz 2014, pp. 61ff.).

Richard Boyd (1984, p. 52)—a supporter of scientific realism—outlines the central arguments of constructivism:

"Roughly, the constructivist antirealist reasons as follows: The actual methodology of science is profoundly theory-dependent. What scientists count as an acceptable theory, what they count as an observation, which experiments they take to be well designed, which measurement procedures they consider legitimate, what problems they seek to solve, and what sorts of evidence they require before accepting a theory—which are all features of scientific methodology—are in practice determined by the theoretical tradition within which scientists work. What sort of world must there be, the constructivist asks, for this sort of theory-dependent methodology to constitute a vehicle for gaining knowledge? The answer, according to the constructivist, is that the world that scientists study, in some robust sense must be defined or constituted or 'constructed' from the theoretical tradition in which the scientific community in question works. If the world that scientists study were not partly constituted by their theoretical tradition, then, so the argument goes, there would be no way of explaining why the theory-dependent methods that scientists use are a way of finding out what is true."

The philosophy of science debates within marketing in the 1980s and 1990s were dominated by the conflict between followers of relativism and scientific realism (for an overview see Hunt 2014; Kavanagh 1994). Now it looks like the "career" of relativism in marketing research is over. One logical problem with relativism might be influential in this context: relativists argue that *all scientific* knowledge is dominated by context factors. If relativism is a *scientific* position this should be true for relativism itself and this would mean that the relativistic philosophy of science is heavily influenced by its context (interests, influences, etc.) as well. See Fig. 3.1.

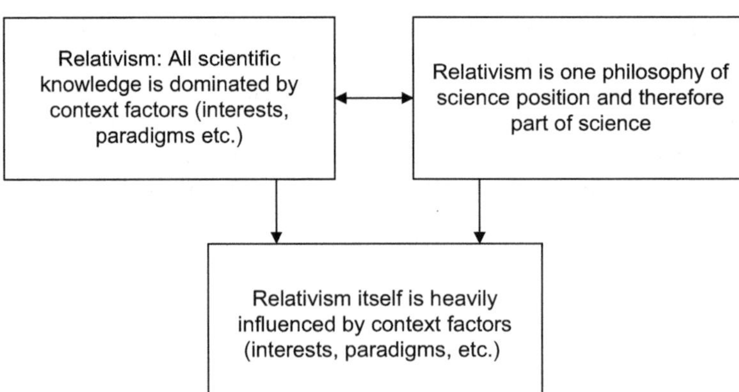

Fig. 3.1 Logical problem of relativism

Ronald Giere (1999, p. 20) characterizes a logical problem of relativism/constructivism:

"What is the character of constructivist sociology of science itself? The problem becomes a reflexive problem if one answers that the sociology of science is itself a science, subject to the same sorts of investigation constructivists have carried out for other sciences. Epistemological constructivists would have to conclude that their own beliefs about the scientists they study were determined more by their own interests and social interactions as sociologists than by whatever might really be going on among their subjects. For ontological constructivists, the results of such an investigation would have to be that the objects of investigation, the beliefs, interests, etc., of their subject scientists are constituted by their own practices as constructivist sociologists."

We now proceed with the *characterization of scientific realism* (SR), which forms the conceptual basis for essential parts of this book. SR agrees with the critical rationalism of Karl Popper concerning the fallibilism of scientific knowledge and realism, but takes a counter-position to the rejection of inductive reasoning. For this purpose, SR assumes that the multiple empirical confirmations of a theory speak for its (approximate) truth (see below), which is particularly clearly formulated in the inductive-realistic model of Hunt (2010, 2011) (see Sect. 5.3).

Important for the establishment of SR is the aspect of the long lasting success of science(s) and its implications. Over about 400–500 years numerous scientific discoveries—with all their imperfections and inconsistencies—have proven themselves many times and thus have provided evidence for their closeness to truth. From the vast abundance of examples, let us mention just a few:

- In medicine, we have learned about many different infections—how they develop and how to prevent them. As a result, numerous health risks have been dramatically reduced and some diseases (almost) eradicated.
- Engineers and architects have such accurate and solid knowledge of statics, properties of materials, etc. that buildings hardly ever collapse even in (seemingly) boldly constructed structures.
- In astronomy and physics, the understanding of the attractions of celestial bodies has become so comprehensive and accurate that in 2014, the spacecraft "Rosetta" after 10 years (!) of flight was able to circle the sun many times and to land on a comet not bigger than 4 km wide.

In marketing, too, despite some deficits compared to the (much older) natural sciences, a considerable body of knowledge has developed over the decades (see Eisend 2015). Here are some examples:

- Certain forms of compensation of sellers have corresponding effects on their behavior.
- Advertising effects are known to the extent that advertising campaigns can be designed and implemented with considerable likelihood of success.
- With proper sampling and implementation, market research can reasonably infer from a small sample of consumers to the consumer behavior in a particular market.

Would these and countless other successful applications of scientific knowledge (not least in medicine and technology) be plausible, if one had to assume that science was essentially subjective or influenced by social conditions and a systematic approach to truth could not be expected? Hardly likely. The centuries-long success of modern science would be a *miracle*. In relativistic terms, how do you explain that astronauts who flew millions of miles through space actually returned to Earth, or that a small amount of a vaccine actually prevents anyone from getting polio? All this is only plausible if one assumes (not relativistically) that scientific research leads to an *approximation* to a (not completely known) true understanding of reality. This is called the "**no-miracles argument**" or sometimes the "success of science argument" (Devitt 2008, p. 227). Smart (1963, p. 39) argues that the success of science over the centuries would have to be based on a huge number of "cosmic coincidences" if scientific statements had no correspondence to reality. This consideration is of central importance for the foundation and justification of realism: "The only reasonable explanation for the success of theories of which I am aware is that well-confirmed theories are conjunctions of well-confirmed, genuine statements and that the entities, to which they refer, in all probability exist" (Maxwell 1962, p. 18).

Scientific realism is thus in clear opposition to relativism (Hunt 1990, 2010). While essential aspects of critical rationalism (such as fallibilism and realism) are compatible with SR, the relativists' typical view of a science that is essentially determined by the context of its development is fundamentally inconsistent with SR's central ideas.

On the one hand, the "no-miracles argument" is convincing and intuitively easy to understand, and on the other hand, in contrast to relativism, it is an essential basis for SR. Nevertheless, there is also a considerable objection to the centuries-long conclusion of scientific successes, that realism is the only explanation for the success of science. Realism may indeed (currently) be the "best explanation" for it, but it is not a logically compelling conclusion (Schurz 2014, p. 294; Lipton 2008). Other explanations cannot be completely ruled out, even if they are not easily imaginable at present. Although the adequacy of SR is not "proven", it is considered by many (probably the majority of) scientists—including the authors of this book—to be the most appropriate of the current science-based approaches.

Stathis Psillos (2007, p. 166) characterizes the relevance of the no-miracles argument:

"No matter how exactly the argument is formulated, its thrust is that the success of scientific theories, and especially their ability to issue in novel predictions, lends credence to the following two theses: (1) that scientific theories should be interpreted realistically; and (2) that, so interpreted, these theories are approximately true. On a realist understanding of theories, novel predictions and genuine empirical success is to be expected."

Martin Carrier (2004, p. 140) formulates two basic assumptions of SR (emphasis added by the authors of this book):

1. "The theoretical terms in the *mature* sciences typically refer to real objects."
2. "The theoretical laws in the *mature* sciences are typically *approximately true*."

At first, it is necessary to explain what is meant by "**mature sciences**". Carrier does not define this term exactly, but it is illustrated by the examples of physics, chemistry and biology. Commonly, the term "maturity" is used to characterize a high and stable level of development. In the above context, this term denotes sciences that have, over a long period, developed a comprehensive body of knowledge, laws and theories with a relatively high degree of validity.

The concept of **approximate truth** mentioned by Carrier (2004) plays a central role for SR (see also Boyd 2002; Chakravartty 2011). Although one can identify some exceptions in the history of science, SR assumes that theories and statements in "mature" sciences are approximately true, so that typically the deviations from a (probably never achievable) completely certain and precise knowledge are small (or decrease with growing research). Psillos (1999, pp. 276ff.) emphasizes that a complete (not approximate!) correspondence between theoretical statements and reality is hardly possible, on the one hand due to theories that simplify the reality, and on the other hand, because observations or measures of real phenomena are usually flawed. In addition, here arises the logical problem that one can estimate the degree of approximation to a truth only if one knows this truth, which, of course, is usually not the case. If we knew the truth, we wouldn't need approximations. This very simple example (see Psillos 2007, pp. 12–13) may illustrate the character of approximate statements: The statement "Alfred is 1.760 meters tall" is false when Alfred is actually 1.761 meters tall; however, the statement is at least approximately true. In most cases, such an accuracy level is sufficient in marketing. Who needs to know whether a market share is exactly 20% or 20.1% or market growth is 2.08% or 2.11%? In scientific research, in many cases it is more relevant to know whether a difference or a correlation is "significant" (see Chap. 7) than the exact numbers. A well-known example from the natural sciences is Newton's law of gravitation. This was challenged by Einstein's "Theory of Relativity", but was still used for calculations because the differences in the results are minimal. Weston (1992,

p. 55) proposes a formulation that simultaneously corresponds to the approximate character of statements and the typical fallibilism of SR: "Available evidence *indicates* that the theory is *approximately* true".

Accordingly, some of the limitations that have already been mentioned in regard to induction, or the problem of "pessimistic induction" (see Sect. 3.2), no longer arise with full sharpness. The concept of approximate truth has repeatedly been critically discussed and questioned with regard to its concretization. Shelby Hunt (2011, p. 169) has developed the following labeling in connection with his "inductive-realistic" model: "Accepting a theory (. . .) as approximately true is warranted when the evidence related to the theory is sufficient to give reason to believe that something like the specific entities, the attributes of the entities, and the relationships, structures, and mechanisms posited by the theory is likely to exist in the world external to the theory".

Now returning to the central characteristics of SR: After his—in part sharp—criticism of the temporary (in the last third of the twentieth century) influential relativism, Shelby Hunt formulated (e.g., 1990; 2003, pp. 170ff; 2010, pp. 225ff, Hunt and Hansen 2010), on the basis of the relevant philosophical literature (see the overviews in Boyd 2002; Hunt 2010), a *concept of scientific realism*, which is, in his view, a much better alternative. Hunt characterizes this concept in four tenets, which are briefly explained here:

- **"Classical" realism:**

It is assumed that a reality exists, independent of the perception and view of the observer. Psillos (2006, p. 688) speaks of the "metaphysical thesis" of scientific realism in the sense that the corresponding statement ("The world has a definite and mind-independent structure") lies beyond the realm of experience and is not confirmed or even proven, but rather "believed"—or not. In this thesis, SR fully agrees with critical rationalism and differs significantly from relativism.

Michael Devitt (2008, p. 225) distinguishes two aspects of realism:

"Common-sense realism: Most of the observable physical entities of common sense and science exist mind-independently.

Scientific realism: Most of the essential unobservable of well-established current scientific theories exist mind-independently."

Richard Boyd (1984, p. 42) summarizes: "The reality which scientific theories describe is largely independent of our thoughts or theoretical commitments".

- **"Inductive" realism:**

If a theory and its statements are confirmed in the long run and in many appropriate tests and practical applications, then there are evidently many indications that these statements are, with relatively high probability, approximately correct, although of course no full certainty can be achieved. Obviously, a great number of corresponding empirical findings strengthens the trust in a scientific statement. By abandoning the "fundamentalist claim" (see above) of the certainty of knowledge, the central objection to inductive reasoning becomes ineffective. Thus, in the case of inductive conclusions, some degree of uncertainty of the statements is accepted. This leads to the subsequent aspect of fallibility.

Ernan McMullin (1984, p. 26) summarizes the central idea of "inductive realism" in one sentence: "The basic claim made by scientific realism (...) is that the long-term success of a scientific theory gives reason to believe that something like the entities and structure postulated by the theory actually exists."

- **"Fallibilistic" realism:**

Complete certainty that knowledge of reality is correct (for example in the sense of a logically definite conclusion) cannot be achieved. Again, SR is compliant with the viewpoint of critical rationalism (see above) that scientific statements are typically fallible ("fallibilism"), and the history of science has shown many cases of scientific statements or theories—accepted in earlier times—as being wrong. Even with a *relativistic* worldview, there is some agreement, because this view emphasizes the uncertainty of scientific statements because of their assumed contextual dependence. Proponents of SR certainly accept that the emergence of scientific knowledge may be *influenced* by political, intellectual, economic, etc. framework conditions (see Sect. 3.2), but do not (as proponents of relativism) assume that scientific knowledge is *mainly determined* by the particular context.

Gerhard Schurz (2014, p. 23) briefly characterizes the core of fallibilism: "According to the assumption of fallibilism, every scientific statement is more or less fallible; so we can never be absolutely sure of their truth, but we can consider their truth to be more or less probable."

- **"Critical" realism:**

One of the central tasks of science is to question statements about reality in terms of their correctness and to gain knowledge that best suits reality. From the fallibilism mentioned above (and in Sect. 1.2) and from the limitations of inductive reasoning (see Sect. 2.5), the task of gaining "better" knowledge arises. Continuous critical questioning and new research play a central role in this task. This refers to the idea of critical rationalism, whose name already contains this idea of *continual critical questioning*.

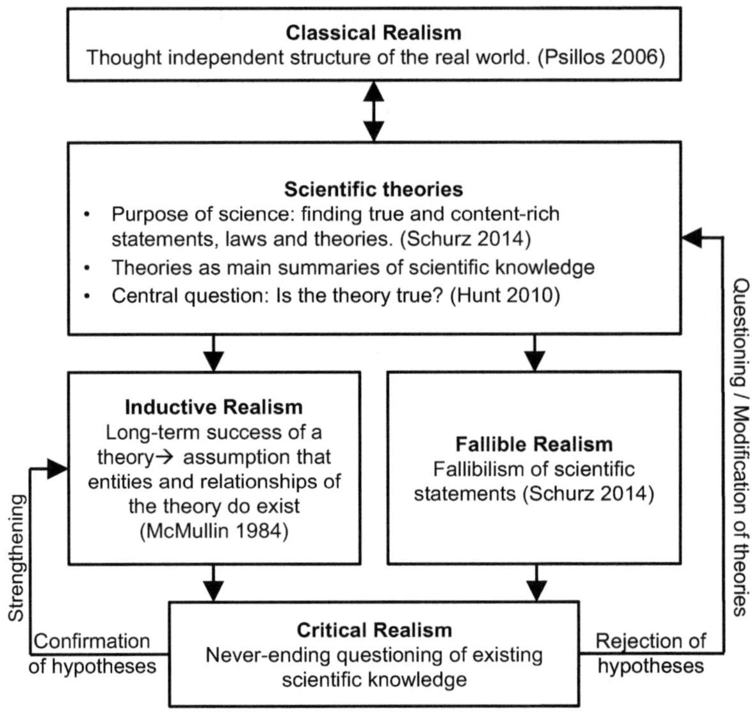

Fig. 3.2 Characteristics of scientific realism

Gerhard Schurz (2014, p. 23) briefly and clearly formulates the crucial idea that "everything depends on using empirical tests to find out more about the probability of a scientific hypothesis (. . .). Fallibility goes hand in hand with a critical attitude, according to which no statement may ever be once and for all exempted from criticism".

Figure 3.2 presents the four principles of scientific realism and essential relationships in a simple summary. It should be emphasized that this presentation is not about processes (such as a "research process"), but about connections between ideas. The double-sided arrow shown in Fig. 3.2 between "the real world" (that exists independent of perception → "*classical realism*") and corresponding theories is intended to indicate that in *mature* sciences one can usually assume an approximate agreement (or "approximate truth"). On the one hand, the evaluation of a theory raises the question of the extent to which it is confirmed in empirical tests and applications ("*inductive realism*"). On the other hand, the principle of fallibilism is that a previously accepted theory may turn out to be false ("*fallible realism*"). Both aspects suggest that a theory should be repeatedly critically examined—usually empirically ("*critical realism*"). Depending on the results of such tests (the confirmation or rejection of hypotheses), the result is an increase in the acceptance of the theory or the questioning of the theory or its modification. This point of view is discussed again in the context of the inductive-realistic model in Sect. 5.3.

Richard Boyd (2002, p. 1) briefly summarizes key ideas of SR:

"Scientific realists hold that the characteristic product of successful scientific research is knowledge of largely theory-independent phenomena and that such knowledge is possible (indeed actual) even in those cases in which the relevant phenomena are not, in any non-question-begging sense, observable. According to scientific realists, for example, if you obtain a good contemporary chemistry textbook you will have good reason to believe (because the scientists whose work the book reports had good scientific evidence for) the (approximate) truth of the claims it contains about the existence and properties of atoms, molecules, sub-atomic particles, energy levels, reaction mechanisms, etc. Moreover, you have good reason to think that such phenomena have the properties attributed to them in the textbook independently of our theoretical conceptions in chemistry. Scientific realism is thus the common sense (or common science) conception that, subject to a recognition that scientific methods are fallible and that most scientific knowledge is approximate, we are justified in accepting the most secure findings of scientists 'at face value'".

For marketing research, SR has consequences that have become almost a matter of course. For example, the concept of attitude that has been successfully applied over decades—in social science research and in practice—is widely accepted. The long-term success of attitude theory speaks in many examples of the existence of the relevant concepts (e.g. attitude) and related structures (e.g. the relationship between attitude and behavior). Furthermore, the empirical methodology mainly used in empirical marketing research (see Chaps. 6, 7, 8 and 9) largely corresponds to central ideas of SR (→ inductive realism), because reviews of theories based on relatively large samples are essential for decisions concerning the acceptance or rejection of hypotheses (and theories). Ultimately, it can be observed that orientation along SR contributes to trust in corresponding research results (Hunt 2010) because, in this sense, relatively broadly shared knowledge is generated, rather than more or less subjective ("relativistic") representations of perceptions and appraisals of reality. This leads to the acceptance of marketing research statements by other scientists, both inside and outside the discipline, as well as among students and practitioners. What relevance would academic teaching or expert opinions have, if they were based on a largely subjective understanding of science? At least in the context of justification (see Sect. 1.2), one could not do justice to the goal of the objectivity of statements.

3.2 Criticism of Scientific Realism

So far, SR as the basis of many of the reflections in this book has been characterized by its central features. In the following section, some further thoughts will follow. It has already become apparent that SR clearly differs from other philosophy of science approaches. Thus, it is not surprising that a critical discussion is taking place in the philosophy of science. Some major points of criticism are briefly outlined in this section:

- *Underdetermination* of theories (ambiguity in the relationship between theory and empirical results)
- "*Pessimistic induction*" (inference from negative experiences in the history of science to the evaluation of current theories)
- Influence of social or historical *context* on scientific knowledge and theory-ladenness.

Underdetermination of theories

Underdetermination refers to the problem that some observations or constellations of data allow different theoretical interpretations and thus cannot confirm the truth of a *particular* (single) theory. "All underdetermination arguments exploit the fact that often more than one theory, explanation or law is compatible with the evidence" (Ladyman 2002, p. 162). It may well be that the same results of observations are compatible with several theories. When it comes to the existence of alternative (already known or not yet known) theories that can explain certain observations in different ways, it is impossible to determine which of these theories is (approximately) true (Psillos 1999, p. 162).

One can distinguish two types of underdetermination (see Stanford 2013), which are referred to here (with regard to the topic of this book) in somewhat narrower terms than in Stanford (2013, p. 2):

- **"Alternative underdetermination"** refers to the cases in which observations are compatible with different theories, in which the observations do not speak clearly for a particular theory and the truth of alternative theories is not clear (see above).
- **"Measurement error underdetermination"** characterizes the problem that the confirmation (or questioning) of a theory after an empirical investigation and hypothesis test does not necessarily have to be determined by the (possibly lacking) agreement of the theory with the corresponding parts of reality. Rather, it may well be that such a result is due to errors (usually not completely avoidable) appearing in the process of empirical research (problem of validity).

An illustrative example of *alternative underdetermination* is the so-called "curve fitting problem" illustrated in Fig. 3.3 (see also Phillips and Burbules 2000, pp. 17ff.; Newton-Smith 2000). Clearly certain sets of measured values allow different interpretations (→ theories), in this simple example linear and one (or many)

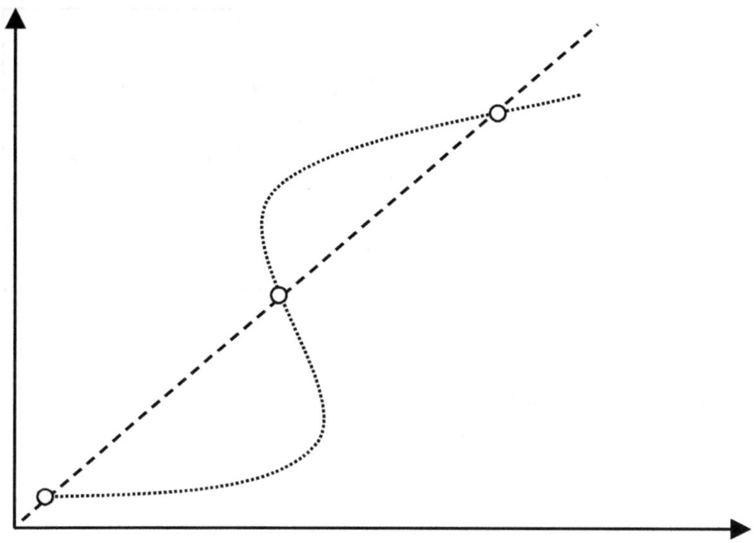

Fig. 3.3 Example for underdetermination of theories through data (following Phillips and Burbules 2000, p. 18)

non-linear relationships. "If we consider any finite group of data points, an elementary proof reveals that there are an infinite number of distinct mathematical functions describing different curves that will pass through all of them." (Stanford 2013, p. 10).

This creates the problem of "empirical equivalence", i.e. that of "alternative theories making the very same empirical predictions, and which therefore cannot be better or worse supported by any possible body of evidence" (Stanford 2013, p. 11). One may wonder whether individual examples suffice to speak of a *general* problem of empirical equivalence. Often, the problem of alternative theories is more likely to be hypothetical; researchers are often satisfied if they find at least one plausible theory that corresponds to the present observations. Although alternative theories do exist, they are unlikely to be *equivalent* (Okasha 2002, pp. 72–73) because empirical support of a theory, while certainly important, is *not the only* relevant criterion (see Sect. 5.1). Figure 3.4 illustrates the problem of alternative underdetermination with an example in which an empirical finding corresponds to two different theories simultaneously.

The problem that empirical data are typically flawed (due to measurement errors, sampling errors, etc.) also exists from the perspective of SR. **Measurement error underdetermination** thus refers to the problem that the rejection or acceptance of a scientific hypothesis does not necessarily have to be based on the falsity or correctness of the corresponding theory. It may therefore well be that the rejection (or confirmation) of a theory would be overhasty because the reason for not confirming (or confirming) a hypothesis may be due to shortcomings of the empirical

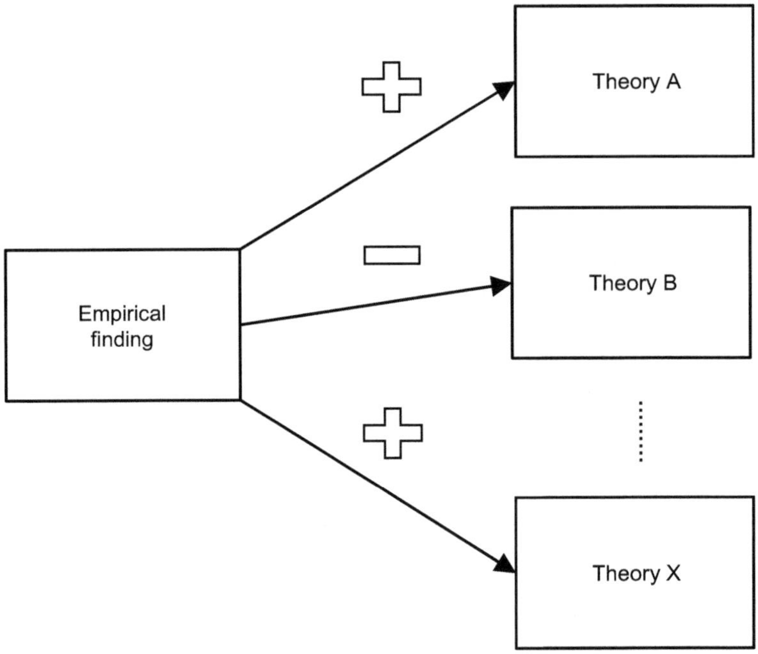

Fig. 3.4 Correspondence of an empirical finding to alternative theories (alternative underdetermination)

investigation. This problem corresponds to the so-called Duhem thesis (Duhem 1906/1954), which states that the predictions of a theory are bound to the validity of accompanying assumptions, e.g. measurement properties or methods. "If the prediction is not fulfilled, the only thing we can logically infer is that *either* the auxiliaries *or* the theory is false" (Psillos 2007, p. 71); see also Fig. 3.5. Strictly speaking, the possibility that one cannot say with certainty whether a hypothesis is wrong or that this hypothesis was rejected because of measurement errors, would mean that a statement is *ultimately* not falsifiable.

Pierre Duhem (1906/1954, quoted in Curd and Cover 1998, p. 263) has summarized his proposition:

"In sum, the physicist can never subject an isolated hypothesis to experimental test, but only a whole group of hypotheses; when the experiment is in disagreement with his predictions, what he learns is that at least one of the hypotheses constituting this group is unacceptable and ought to be modified; but the experiment does not designate which one should be changed."

The most important approach to influence this problem is the development and application of appropriate methods, regardless of the philosophy of science position

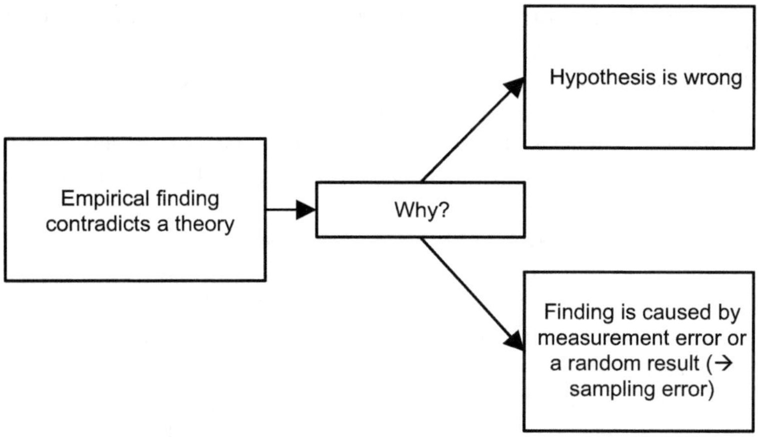

Fig. 3.5 Illustration of measurement error underdetermination

of the researcher. ***Reliability***, and particularly ***validity***, of measurements (see Chap. 6) therefore play a central role in empirical marketing research. Scientific realism adds a point of view that does not solve the problem but reduces it. This happens due to the typical assumption of SR that a *multitude* of empirical results confirming a theory suggests (but does not prove!) that this theory is (close to) truth ("inductive realism"). Thus, if a large number of such test results are the basis for the acceptance of theories, then in many cases one can assume that these results are not all error-prone in the "same direction", but that deviations, due to measurement errors, compensate each other to a certain extent. This is especially true when various investigations have been carried out by different researchers, in different contexts (i.e. under different social, cultural and economic conditions) and with different methods. Please refer to the discussion of meta-analysis in Chap. 9.

Which *problems* result from the two facets of underdetermination for SR? First, the inductive reasoning of SR is called into question: If one finds empirical confirmations for one theory, then it could well be that these results also confirm another (perhaps as yet unknown) theory (→ alternative underdetermination). Measurement error underdetermination leads to doubts about the meaningfulness of empirical results: faulty empirical studies limit the possibilities of inferring from their results that the corresponding theory or hypothesis is true or not (→ Duhem's thesis).

So, if the confirmation of a theory by empirical results cannot be clearly determined, then the question arises, what other factors could affect the acceptance of a theory? Perhaps, as some relativists (see Sect. 3.1) believe, the context may play the decisive role. This can certainly not be *concluded* from underdetermination, because this does not say anything about the factors influencing the decision to accept a theory (Stanford 2013). The aspect of the underdetermination of theories leaves it completely open whether, for example, social or completely different influences determine the choice of a theory. In addition, SR argues that the acceptance of a theory does not depend on a single empirical verification, but rather on a larger number of "empirical successes" (see the inductive-realistic model in Sect. 5.3). As the number of studies increases, dependence on the character of individual researchers tends to decrease. In addition, the growth and evolution of empirical research methods (such as the use of standardized measurements) lead to inter-subjectively more predictable results and fewer opportunities for manipulation.

Pessimistic induction (history of science experience)
This point refers to the experience that in the past, some successful theories have later been shown to be wrong (e.g., Devitt 2008, pp. 232–233). Larry Laudan (1981), for example, has compiled a list of natural science theories that seemed to be temporarily accepted and well endorsed, but were later rejected. In marketing research, one may still remember the microeconomic price theory or the AIDA model of advertising effects, both theoretical approaches that have no relevance today. The *argument* of "pessimistic induction" here refers to an inductive conclusion of (partially negative) experiences with earlier theories on the assessment of contemporary theories: if earlier theories have been confirmed for only a certain time, and not permanently, then even with today's (apparently) well-confirmed theories, one must expect that new theories will endure only partially. However, this argument would only be relevant in general if, in fact, the failure of previously accepted theories occurred frequently or was very common in science.

Stathis Psillos (1999, p. 101) characterizes the "pessimistic induction" critique in regard to SR:

"Laudan's argument against scientific realism is simple but powerful. It can be summarized as follows: The history of science is full of theories at different times and for long periods had been empirically successful, and yet were shown to be false in the deep-structure claims they made about the world. It is similarly full of theoretical terms featuring in successful theories which do not refer. Therefore, by a simple (meta-) induction on scientific theories, our current successful theories are likely to be false (or, at any rate, are more likely to be false than true), and many or most of the theoretical terms featuring in them will turn out to be non-referential. Therefore, the empirical success of a theory provides no warrant for the claim that the theory is approximately true."

Devitt (2011) repeatedly criticized the argument of pessimistic induction and in this way defended the approach of realism. First, he emphasized that more up-to-date theories are more successful than older theories, and therefore one cannot directly infer from past failures in the history of science up to the present. Related to that, it should be noted that SR refers to *mature* sciences (see Sect. 3.1). This goes hand in hand with the argument that research methods have evolved and improved over time. Failures from past decades or centuries cannot continue to determine expectations of the success of contemporary science.

Michael Devitt (2011, p. 290) summarizes his position briefly:

"Improvements in scientific methodologies make it much harder to mount a case against realism than seems to have been appreciated. For the appeal to historical details has to show not only that we were nearly always wrong in our unobservable posits but that, despite methodological improvements, we have not been getting significantly righter. It seems to me most unlikely that this case can be made."

The problem of the pessimistic induction arises only against the background of a "fundamentalist" view that science has to make *secure* statements (Phillips and Burbules 2000, pp. 5ff.; Schurz 2014, p. 3). This claim may have been applied to some scientific disciplines for a long time but, at least in empirical marketing research, it has been less important. The methods of inferential statistics and the possibility of errors in measurements made it evident that research results and theories are always subject to uncertainty and/or errors. The context of SR makes it clear that, in general, a "fundamentalist" claim for science is no longer made, but the fallibilism of scientific knowledge is assumed (see Sects. 1.2 and 3.1). Thus, a limited number of theories accepted in earlier times, which later turned out to be wrong, do not really challenge the approach of SR in marketing research.

Influence of insights due to the social/historical context and theory-ladenness
In the history of science, one finds examples of changing "world views" that resulted in very different theoretical ideas about relevant parts of reality. This has already been discussed in Sect. 3.1 in relation to relativism. In connection to this there are also widespread social influences on scientists as well as their dependency on research funding. Due to the peculiar reputation and credibility of science mentioned in Sect. 1.1, there are always attempts to influence scientific research in line with certain interests. Here are some examples:

- For centuries, one could observe attempts by religious organizations to influence scientists, just as in the US today so-called creationists seek to replace the Darwinian theory of evolution in school education with the biblical story of creation.
- A somewhat bizarre example from the twentieth century is the failed attempt of the Soviet biologist Trofim Lyssenko to develop genetic theories for the creation of a new human type in the ideological interest of the Communist Party.
- Lobbyists seek to lend weight to their views by commissioning (and paying well for) scientific analyses that produce the desired results.
- Nowadays, third-party funds from private or public institutions often play a significant role in science. Here, too, an influence on the focus and results of research is probably not completely avoidable.

The problem of *external influences* on scientific findings certainly exists independently of philosophy of science positions. In contrast to the view of relativists, however, supporters of SR do *not* assume that scientific statements are typically *shaped decisively* by social, political, cultural and economic conditions or by paradigms, while such—more or less limited—influences cannot be definitely ruled out, especially if applied research affects the interests of certain social groups. This aspect is therefore taken explicitly into account in the most recent version of Shelby Hunt's inductive realist model (see Sect. 5.3).

The problem of **theory-ladenness** refers to the fact that the prior knowledge of an observer, his or her theoretical assumptions, hypotheses, etc. typically influence his or her perceptions and interpretations of reality. For example, a comparison of consumers' buying decisions based on observations of salespeople or academic consumer researchers may show that different views are dependent on their theoretical or practical perspectives and respective experiences. In the process of developing, checking and changing theories, these theories may influence the *perception* of real phenomena. Here, the focus is *only on influencing*, but not by a decisive imprint, as relativists would probably assume.

Theory-ladenness should not be confused with the above mentioned *external influences* on science. The former arises almost inevitably and often unnoticeably in the processes of discovery within science, because competent scientists cannot be without prior knowledge, experience, etc. Context factors—as the term indicates—have an *external* influence on the research process. Fig. 3.6 illustrates this difference.

With regard to theory-ladenness of data, one cannot avoid the fact that humans' perception of reality (see Fig. 3.7) is influenced to a certain extent by their prior knowledge and experience. The problem in SR is somewhat "mitigated" by confirming theories via a (large) number of studies and corresponding practical

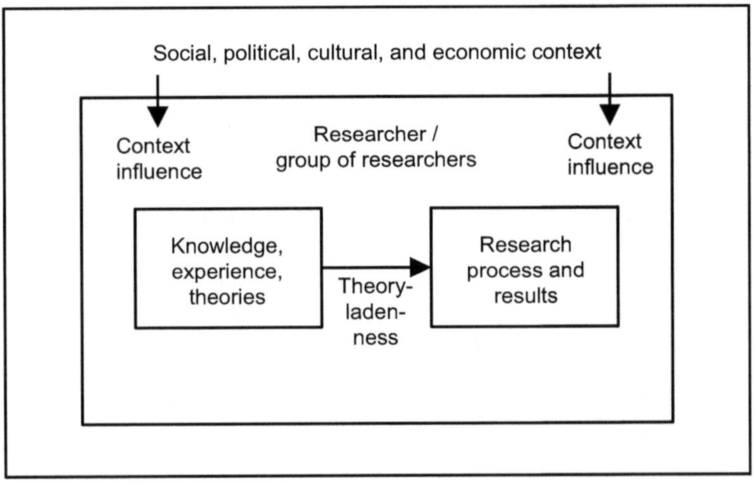

Fig. 3.6 Contextual influences and theory-ladenness

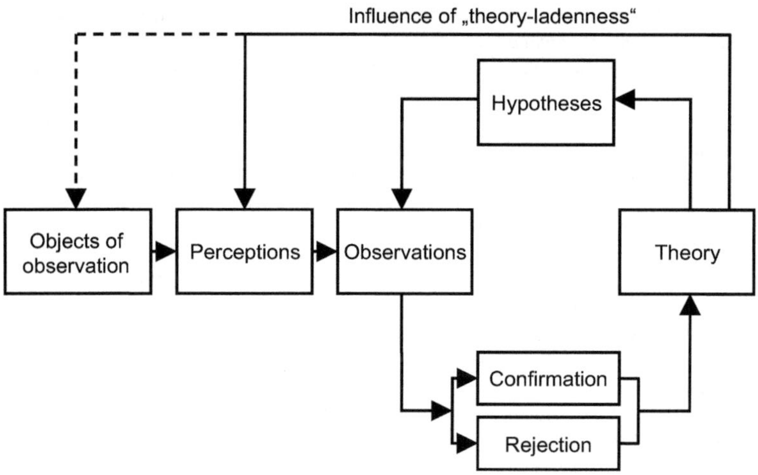

Fig. 3.7 Influence of theory-ladenness (following Hunt 1994, p. 138)

experiences that have been made in the past (see the comments on the inductive realist model in Sect. 5.3). Inevitably, this involves a greater number of researchers and scientists, different points in time and situations. This leads, to some extent, to a diminished dependence on the statements on the theory-ladenness of an *individual* researcher at a particular time and in a particular situation.

References

Baghramian, M. (2008). Relativism about science. In S. Psillos & M. Curd (Eds.), *The Routledge companion to philosophy of science* (pp. 236–247). London: Routledge.

Boyd, R. (1984). The current status of scientific realism. In J. Leplin (Ed.), *Scientific realism* (pp. 41–82). London: University of California Press.

Boyd, R. (2002). Scientific realism. In E. Zalta (Ed.), *The Stanford Encyclopedia of philosophy*. http://plato.stanford.edu

Brown, J. (Ed.). (2012). *Philosophy of science: The key thinkers*. New York: Continuum.

Buzzell, R. D., & Gale, B. T. (1987). *The PIMS principles. Linking strategy to performance*. New York: The Free Press.

Carrier, M. (2004). Experimental success and the revelation of reality: The miracle argument for scientific realism. In M. Carrier (Ed.), *Knowledge and the world: Challenges beyond the science wars* (pp. 137–161). Heidelberg: Springer.

Carrier, M. (2012). Historical approaches: Kuhn, Lakatos and Feyerabend. In J. Brown (Ed.), *Philosophy of science: The key thinkers* (pp. 132–151). New York: Continuum.

Chakravartty, A. (2011). Scientific realism. In E. Zalta (Ed.), *The Stanford Encyclopedia of philosophy*. http://plato.stanford.edu

Chalmers, A. (2013). *What is this thing called science?* (4th ed.). Indianapolis, IN: Hackett.

Curd, M., & Cover, J. (Eds.). (1998). *Philosophy of science—The central issues*. New York/London: Norton.

Devitt, M. (2008). Realism/anti-realism. In S. Psillos & M. Curd (Eds.), *The Routledge companion to philosophy of science* (pp. 285–293). London: Routledge.

Devitt, M. (2011). Are unconceived alternatives a problem for scientific realism? *Journal for General Philosophy of Science, 42*, 285–293.
Duhem, P. (1906/1954). *The aim and structure of physical theory*. Princeton: Princeton University Press.
Eisend, M. (2015). Have we progressed marketing knowledge? A meta-meta-analysis of effect sizes in marketing research. *Journal of Marketing, 79* (3), 23–40.
Giere, R. (1999). *Science without Laws*. In Chicago/London: University of Chicago Press.
Godfrey-Smith, P. (2003). *Theory and reality—An introduction to the philosophy of science*. Chicago/London: University of Chicago Press.
Haig, B. (2013). The philosophy of quantitative methods. In T. Little (Ed.), *The Oxford handbook of quantitative methods* (Vol. 1, pp. 7–31). Oxford: Oxford University Press.
Hunt, S. D. (1990). Truth in marketing theory and research. *Journal of Marketing, 54*(July), 1–15.
Hunt, S. D. (1994). A realist theory of empirical testing: Resolving the theory ladenness/objectivity debate. *Philosophy of the Social Sciences, 24*(2), 133–158.
Hunt, S. D. (2003). *Controversy in marketing theory*. London: Sharpe.
Hunt, S. D. (2010). *Marketing theory—Foundations, controversy, strategy, resource-advantage theory*. London: Sharpe.
Hunt, S. D. (2011). Theory status, inductive realism, and approximate truth: No miracles, no charades. *International Studies in the Philosophy of Science, 25*, 159–178.
Hunt, S. D. (2014). Understanding Marketing's philosophy debates. *Journal of Historical Research in Marketing, 6*(3), 351–378.
Hunt, S. D., & Hansen, J. (2010). The philosophical foundations of marketing research: For scientific realism and truth. In P. Maclaran, M. Saren, B. Stern, & M. Tadajewski (Eds.), *The SAGE handbook of marketing theory* (pp. 111–126). Los Angeles: Sage.
Kavanagh, D. (1994). Hunt versus Anderson: Round 16. *European Journal of Marketing, 28*(3), 26–41.
Kuhn, T. (1970). *The structure of scientific revolutions* (2nd ed.). Chicago/London: University of Chicago Press.
Kuhn, T. (2012). *The structure of scientific revolutions* (4th ed.). Chicago/London: University of Chicago Press.
Ladyman, J. (2002). *Understanding philosophy of science*. London: Routledge.
Laudan, L. (1981). A confutation of convergent realism. *Philosophy of Science, 48*, 19–49.
Leplin, J. (Ed.). (1984). *Scientific realism*. Berkeley: University of California Press.
Lipton, P. (2008). Inference to the best explanation. In S. Psillos & M. Curd (Eds.), *The Routledge companion to philosophy of science* (pp. 193–202). London: Routledge.
Maxwell, G. (1962). The ontological status of theoretical entities. In H. Feigl & G. Maxwell (Eds.), *Scientific explanation, space and time. Minnesota studies in the philosophy of science* (Vol. 3, pp. 3–27). Minneapolis: University of Minnesota Press.
McMullin, E. (1984). A case for scientific realism. In J. Leplin (Ed.), *Scientific realism* (pp. 8–40). Berkeley: University of California Press.
Newton-Smith, W. (2000). Hume. In W. Newton-Smith (Ed.), *A companion to the philosophy of science* (pp. 165–168). Oxford: Blackwell.
Okasha, S. (2002). *Philosophy of science—A very short introduction*. Oxford/New York: Oxford University Press.
Phillips, D., & Burbules, N. (2000). *Postpositivism and educational research*. Oxford: Rowman & Littlefield.
Popper, K. (2002a). *Conjectures and refutations—The growth of scientific knowledge*. London: Routledge.
Popper, K. (2002b). *The logic of scientific discovery*. London: Routledge.
Psillos, S. (1999). *Scientific realism—How science tracks truth*. London: Routledge.
Psillos, S. (2006). Scientific realism. In D. Borchert (Ed.), *Encyclopedia of philosophy* (Vol. 8, 2nd ed., pp. 688–694). Detroit: Macmillan.
Psillos, S. (2007). *Philosophy of science A–Z*. Edinburgh: Edinburgh University Press.

Psillos, S., & Curd, M. (Eds.). (2008). *The Routledge companion to philosophy of science*. London: Routledge.
Putnam, H. (1975). *Mathematics, matter and method*. Cambridge: University of Cambridge Press.
Rorty, R. (2000). Kuhn. In W. Newton-Smith (Ed.), *A companion to the philosophy of science* (pp. 248–258). Oxford: Blackwell.
Sankey, H. (2008). Scientific method. In S. Psillos & M. Curd (Eds.), *The Routledge companion to philosophy of science* (pp. 248–258). London: Routledge.
Schurz, G. (2014). *Philosophy of science. A unified approach*. London: Routledge.
Smart, J. (1963). *Philosophy and scientific realism*. London: Routledge & Kegan Paul.
Stanford, K. (2013). Underdetermination of scientific theory. In E. Zalta (Ed.), *The Stanford Encyclopedia of philosophy*. http://plato.stanford.edu
Swoyer, C. (2003). Relativism. In E. Zalta (Ed.), *The Stanford Encyclopedia of philosophy*. http://plato.stanford.edu
Weston, T. (1992). Approximate truth and scientific realism. *Philosophy of Science, 59*, 53–74.

Further Reading

Bird, A. (1998). *Philosophy of science*. Montreal: McGill-Queen's University Press.
Hunt, S. D. (2010). *Marketing theory—Foundations, controversy, strategy, resource—Advantage theory*. London: Sharpe.
Psillos, S. (2002). *Causation and explanation*. Durham, UK: Acumen.

Theory Building 4

4.1 Conceptualization and Definitions

In the context of characterizations of theories in Sect. 2.1, the meaning of (abstract) concepts and the essential role of concepts for theories has been emphasized. In this respect, there is probably no specific explanation needed that the development of concepts ("conceptualization") is an essential step in the building of theories. "Concepts are the building blocks of theory" (Neuman 2011, p. 62). Closely related to this is the most precise characterization of concepts through corresponding *definitions*, which in turn form the basis for the development of appropriate measurement instruments (see Chap. 6). Against this background, considerations of conceptualization and definitions are an essential step in theory building.

The understanding of the term "**conceptualization**" refers to the process of abstract identification of parts of reality that are of interest, and of summarizing them in terms of thought. In marketing we speak—to give an example—after buying decisions (related to cars, travels, wine etc.), summarizing and abstracting from the individual cases about "customer satisfaction", if expectations before the purchase and experiences after the purchase correspond, or when expectations are exceeded. In this section, the considerations of conceptualization and definitions thus focus on the thoughtful development of individual concepts. The literature (for example, Yadav 2010; MacInnis 2011) also offers broader perspectives on conceptualization, in which the whole process of theory building is labeled as conceptualization.

How can one imagine the process of conceptualization? Deborah MacInnis (2011, p. 140) identifies this process as follows:

> Conceptualization is a process of abstract thinking involving the mental representation of an idea. Conceptualization derives from the Medieval Latin conceptualis and from Late Latin conceptus, which refer to 'a thought; existing only in the mind; separated from embodiment' (. . ..). Thus, conceptualization involves 'seeing' or 'understanding' something abstract, in one's mind.

M. Eisend, A. Kuss, *Research Methodology in Marketing*,
https://doi.org/10.1007/978-3-030-10794-9_4

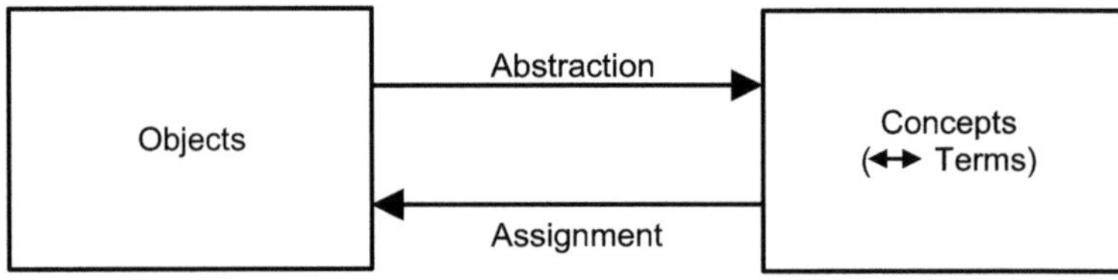

Fig. 4.1 Process of abstraction and assignment (according to Zaltman et al. 1973, p. 28)

If one connects the common features of different objects (people, things, events or states) with a designation that does not refer to individual (specific) objects, but rather to their similarities, neglecting other details—which are of no particular interest—then one *abstracts* from this individual objects (Zaltman et al. 1973, p. 28). For example, people in a hospital are very different in terms of age, gender, ethnicity, occupation, etc.; with regard to a hospital management study, however, it may be necessary to abstract from these features and to talk about "patients". In many cases of scientific research and also practical application, it is essential to assign individual objects to specific concepts. For example, the assignment of a patient to the concept "alcohol-dependent" for his or her treatment and chance of recovery is significant and the assignment of a client to the group of intensive users is important in terms of sales efforts. However, such an assignment can only be successful if the corresponding definition is sufficiently precise. Figure 4.1 illustrates these aspects of conceptualization.

The (mental) development of a concept is often connected with its linguistic characterization, usually by assigning corresponding terms (see Fig. 4.1). This may start with some terms associated with the concept and end with an exact **definition** (see below). The focus is on the process of transcribing a concept. "Instantiation is a deliberate process that involves specifying concrete instances of abstract concepts in order to help clarify their meaning. It is fundamental to science and a crucial process for refining initial theoretical ideas" (Jaccard and Jacoby 2010, p. 76). We could characterize the already mentioned example of customer satisfaction by examples of different types of purchases. This ensures that the relationship between a concept and real phenomena and observations, which is essential for the following empirical tests, remains recognizable.

Of course, if there is sufficient clarity about the content and delineation of a concept, its exact formulation in the form of a definition is required. A definition is the verbal description of a concept and this involves the specification of a mental concept and the possibility of communicating it and making it intersubjective comprehensible. Against this background one also speaks of "**conceptual definitions**". For practical reasons, a written statement is absolutely necessary in order to ensure the necessary precision. "Definition is an operation that introduces a new term on the basis of already existing terms" (Zaltman et al. 1973, p. 26; see also Psillos 2007, p. 62). The new (to be defined) concept is named in the scientific literature as **definiendum**, the defining part of a definition is called **definiens**. For example, Hoyer et al. (2013, p. G-2) define "brand extension" (definiendum) as "using the brand name of a product with a well-developed image on a product in a different category" (definiens).

Shelby Hunt (1987, p. 209) on the nature and usefulness of definitions:

"Definitions are 'rules of replacement' (. . .). That is, a definition means that a word or group of words (the definiens) is proposed to be truth-functionally equivalent to the word being defined (the definiendum). Good definitions exhibit inclusivity, exclusivity, differentiability, clarity, communicability, consistency and parsimony."

"*Inclusivity* means that the phenomena commonly attributed to the definiendum should be included in the definition. By contrast, *exclusivity* refers to the clear distinction from other phenomena."

The way to formulate a conceptual definition is usually anything but easy. It requires appropriate abilities for abstraction, for precise linguistic expression, and for critical reflection. Nevertheless, precise and useful definitions, with regard to the correctness of theoretical statements and corresponding empirical tests, are indispensable. Unclear definitions would not allow a convincing or comprehensible development of theory and formulation of hypotheses. Also, the development of valid measurement instruments is hardly conceivable without a precise definition of the concept (MacKenzie 2003). With a clear focus on research practice, Jaccard and Jacoby (2010, pp. 79ff.) give some advice for common ways to arrive at conceptual definitions:

- Review of extant literature and adoption or modification of existing definitions
- Use of dictionaries and (etymological) dictionaries
- Compilation of essential features of a concept
- Description of the concept in words that are as simple as possible

Definitions of terms are, in principle, free to be chosen. These are only linguistic determinations that do not say anything about reality, insofar as definitions cannot be "right" or "wrong", but only more or less *precise* and *useful*. Essential for this is a largely uniform understanding in the academy community, since otherwise a scientific communication is hardly possible. Here are some "rules" for the formulation of conceptual definitions as Wacker (2004) and MacKenzie (2003) summarize them:

- Definitions should characterize the respective concept as *clearly* as possible and clearly *distinguish* it from other (similar) concepts.
- Definitions should use terms that are as *simple*, *clear* and *concise* as possible.
- Definitions should be *succinct*.
- Definitions should be *compatible* with other definitions in the discipline and previous research.

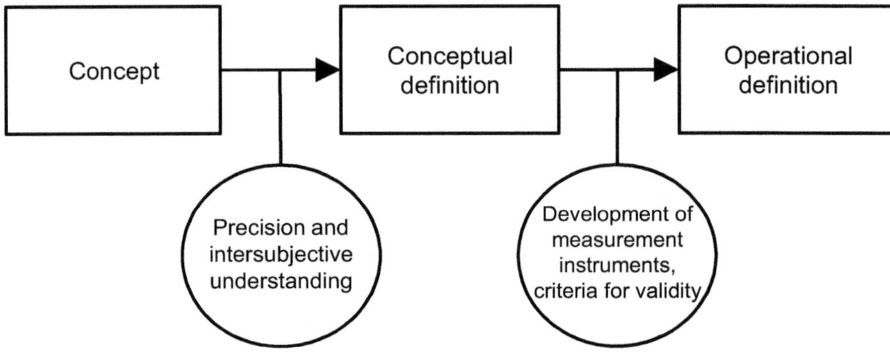

Fig. 4.2 From mental concepts to operationalization

- Empirical studies in which the respective concept plays a role should occur only when the relevant definition has matured to the point that it complies with the above "rules".

With an **operational definition** one goes a step further towards a corresponding measurement for empirical research. "Defining a concept in terms of the instrument or processes used to measure that concept is called 'operationalism' and such definitions are termed operational definitions" (Jacoby and Chestnut 1978, p. 70). We will come back to this process of operationalization and the resulting problems in Chap. 6. Not least, this will be about the correspondence between conceptual and operational definitions. If both (largely) correspond, then one speaks of the **content validity** of a measurement. If there are clear deviations from conceptual and operational definitions, then a corresponding measurement cannot be valid, that is, the result of the measurement has (too) little or nothing at all to do with the concept of interest. Figure 4.2 gives a schematic overview of the steps from the mental concept to the formulation of a conceptual definition to the development of an operational definition, which then allows a corresponding measurement.

4.2 Basic Questions of Theory Building

For decades, the *process* of the emergence of theories in the philosophy of science has received little attention. Some authors (not least Karl Popper) have considered this process to be less structured than it could be and argue that it would be better if the process underwent an analysis by means of psychology, sociology or history of science research. The task of the philosophy of science, from this viewpoint, is concentrated on the following question: "In what sense and to what degree can we trust the results of science?" (Schurz 2014, p. 1). In this context, the distinction between discovery and justification suggested by Hans Reichenbach (1891–1953), which has already been presented in Sect. 1.1, played an essential role. The *context of discovery* is about the development process of theories. Here there exists a wide range of possibilities and hardly fixed rules, as will be seen below. The *context of justification*, on the other hand, refers to rational tests of findings. Discovery contexts were confined to a science-historical interest until the end of the twentieth century, while reasoning and its logic were in the focus of philosophy of science considerations. "The boundary between context of discovery (the de facto thinking processes) and context of justification (the de jure defense of the correctness of these thoughts) was now understood to determine the scope of philosophy of science" (Schickore 2014, p. 6). For details of this development, please refer to Nickles (1985) and Schickore (2014).

A characteristic quote by Karl Popper (2002, pp. 7–8) may illustrate the position of those who do not regard the process of theory formation as an essential subject of philosophy of science:

"I said (. . .) that the work of the scientist consists in putting forward and testing theories. The initial stage, the act of conceiving or inventing a theory, seems to me neither to call for logical analysis nor to be susceptible for it. The question how it happens that a new idea occurs to a man—whether it is a musical theme, a dramatic conflict, or a scientific theory—may be of great interest to empirical psychology; but it is irrelevant to the logical analysis of scientific knowledge. The latter is concerned not with questions of fact (. . .), but only with questions of justification or validity (. . .). Its questions are of the following kind. Can a statement be justified? And if so, how? Is it testable? Is it logically dependent on certain other statements? Or does it perhaps contradict them? In order that a statement may be logically examined in this way, it must already have been presented to us. Someone must have formulated it, and submitted it to logical examination."

It was not until about 1980 that there was a shift in emphasis towards **discovery** contexts, which was primarily initiated by a correspondingly oriented group of philosophers of science (the "*friends of discovery*", Hunt 2013). This is not surprising from today's point of view, because a great number of theoretical and research-related questions arise in relation to scientific discoveries, for example, "Can there be a logic or method of discovery?"; "Must a discovery be both new and true?" (Nickles 2000, p. 85). In addition, there are numerous situations in research practice in which one has to make an effort to build a theory, for example, in the search for explanations for (even practically) relevant phenomena or in the foundation of PhD dissertations. Meanwhile, it is common knowledge that scientific discoveries rarely come about through a sudden single idea ("Eureka!"), but usually it takes longer processes of creation and reviewing. Furthermore, the process of development of a theory is often relevant in regard to credibility (Nickles 2008). In the context of this book, the question of whether scientific discoveries are the subject of philosophy of science is ultimately not really important, because the path to it marks an important task for a researcher.

The term *discovery* is not common in marketing research; it is associated more with the acquisition of knowledge in the natural sciences (for example, certain substances for medical purposes), in astronomy, or (in earlier centuries) in geography. In marketing research, one usually has to deal with theories that have to be developed (often laboriously). Nevertheless, considerations of the context of discovery can be transferred to theory building because, with the development of a (successful) theory, relationships between relevant phenomena are discovered (Hunt 2013).

As has already been noticed, the temporary exclusion of the context of discovery from philosophy of science considerations was also based on the fact that one imagined discoveries as sudden inspirations, the realization of which was hardly comprehensible or even plannable. The experience of extensive work in laboratories, or the processes of theory building, show us that creativity alone is not enough. Rather, the relationship between creativity and the corresponding (empirical) observations and the argumentative justification of the statements and their critical reflection is typical (see Sect. 4.3.2). In this sense, the context of discovery and the context of justification are very often intertwined (Nickles 2008). This experience or perspective is also present in Sect. 4.3 that follows. There, three—by nature very simplified (but common)—ways of theorizing ("Theoretical-in-isolation", Grounded Theory, Empirical Generalizations) are presented.

Section 2.5 presented scientific inferences, which also play an essential role in the development of theories: induction, deduction and abduction. Table 4.1 summarizes the key features of these clauses. Deductive and inductive approaches of theory building (of course) have specific advantages and disadvantages. In **deduction**, existing theories can be linked to corresponding assumptions, concepts and methods, as well as to results obtained in other frameworks (for example, in other scientific disciplines like psychology and consumer behavior), which may increase the efficiency of research. In addition, there is the significant advantage that deduced theories can be relatively well classified in the already existing theoretical inventory. At the same time, this means that completely new perspectives, which might allow a totally different and better understanding of the phenomena of interest, are relatively rare. To that end, **induction** is much more open. Here one begins from the basis of the respective data or experiences to a view corresponding to the respective problem, which are not determined by previous ideas. But this has the disadvantage that theories developed in this way are quite isolated. It should be remembered here (see

Table 4.1 Scientific conclusions at a glance

	Induction	Deduction	Abduction
Basic idea	From many observations to generalization	Derivation of special statements from general statements	Deciding on the most plausible ("best") explanation of a phenomenon
Knowledge development	Expanding knowledge	Truth-preserving	Expanding knowledge
Certainty of conclusions	Uncertain	Certain	Uncertain

Sect. 2.5) that induction is more likely to give rise to hypotheses about laws and lawlike generalizations than to (more complex) theories, which also contain elements that are not observable, and are thus inaccessible to induction (Schurz 2014, p. 53). A frequently used inductive way to generate such hypotheses are empirical generalizations (see Sect. 4.3.4).

Section 2.5 also sketched a third conclusion: **abduction**. These are conclusions from observations on their (assumed) causes. It may be that one makes a selection from a set of well-known relevant hypotheses ("selective abduction") or develops a completely new plausible hypothesis ("creative abduction"). Magnani (2009, see also Schickore 2014) illustrates this with an example from the field of medicine: When a diagnosis is sought for the causes of an illness, the doctor often refers to already known hypotheses about the possible causes of the observed symptoms. In contrast, a creative abduction might be required if it is a new disease, for which there is no experience. Obviously creative abduction leads to more innovative results in theory building than selective abduction.

Even if one does not regard the process of theory building as arbitrary or accidental (see above) and does *not* assume that it usually involves sudden more or less ingenious inspirations, one is, of course, not in a position to have exact rules for this process or to specify "recipes". Therefore the following Sect. 4.3 presents only three different (greatly simplified) approaches to theory building that are quite typical for research practice.

4.3 Approaches to Theory Building

4.3.1 Ideas or Data as a Starting Point?

How can one imagine the emergence or the development of a theory? It already has been suggested that there are no "recipes" or patterns with well-defined procedures (e.g., "steps 1 through n"). If one remembers that there have been several references in this book to empirical testing of existing or proposed theories, then one could get the impression that the first step is in the light of previous experiences, older theories, etc., to make considerations that may/should lead to the design of a new (or modified) theory. This is a process that is specified by the development and use of concepts (see Sect. 4.1), the considerations of relationships between concepts, and the appropriate critical reflections. Ehrenberg (1993) coined the catchy term "**theoretical-in-isolation**" ("TiI") for such an approach. In this view "ideas" are at the beginning of the theory-building process; sometimes, theories from other disciplines are also used, for example, Markov models (Lilien et al. 1992) or approaches from microeconomic theory. Applying more general theories to a particular problem would be a deductive approach (see Sect. 2.5). Section 4.3.2 shows a form of theory formation characterized largely by mental processes.

There is a completely different way of building a theory in research practice, which has been practiced successfully for centuries, especially in the natural sciences. In this type of process observations (e.g., the course of planets in the solar system or growth conditions of certain plants) and the resulting *data* are at the beginning. Based on this, one looks for explanations for these phenomena and builds corresponding theories. These observations are achieved by recording the corresponding natural processes (e.g., astronomy). But there are also countless examples of a different approach. In relevant experiments, phenomena of interest are, so to speak, "generated" in order to be able to make corresponding observations. It is important at this stage that this application of experiments *differs* significantly from the usual approach in marketing research where experiments are conceived of as a particularly rigorous *form of theory testing* (see Chap. 8).

Ian Hacking (1982, p. 71 f.) on the role of experiments in physics:

"Different sciences at different times exhibit different relationships between 'theory' and 'experiment'. One chief role of experiment is the creation of phenomena. Experimenters bring into being phenomena that do not naturally exist in a pure state. These phenomena are the touchstones of physics, the keys to nature and the source of much modern technology. Many are what physicists after the 1870s began to call 'effects': the photo-electric effect, the Compton effect, and so forth."

Why this reference to the role of experiments in other disciplines? It illustrates that the empirical extraction of data can also be the beginning of the process of theory building. In marketing research this is present in two forms: explorative (pre-) studies using qualitative methods (see Sect. 4.3.3) and empirical generalizations (see Sect. 4.3.4). In the latter case, Ehrenberg (1993) identifies the process of theory development on the basis of corresponding results with the term "**empirical-then-theoretical**" ("EtT").

The different approaches also relate to more fundamental considerations about the process of theory building and testing (see Ehrenberg 1993; Hubbard and Lindsay 2013).

- **"Theoretical-in-Isolation" (TiI)**

The building of theory, shaped by ideas and cognitive processes, with subsequent empirical testing, has been established in marketing research for decades. In this way, a theory orientation of research is guaranteed and an unsystematic collection of any data with the publication of incoherent—sometimes rather random—results can be avoided. However, there are doubts as to whether realistic, empirically successful and enduring findings emerge in this way (Ehrenberg 1993).

Very common in marketing research is the use of the hypothetical-deductive method (see Sect. 5.2), in which hypotheses are derived from theoretical statements, whose confirmation or non-confirmation are the decisive criteria for the evaluation of the developed theory. However, the appropriateness of the hypothetical-deductive method is not unlimited (see Sect. 5.2). There are also increasing doubts about the meaning of the commonly used significance tests (see Chap. 7).

- **"Empirical-then-Theoretical" (EtT)**

Here, various (quantitative) empirical studies, the results of which are summarized in corresponding *empirical generalizations* (see Sect. 4.3.4), are the starting point. Ehrenberg (1993, p. 80) recommends: "Develop (low-level) theoretical model or explanation". Inductive and abductive inferences (see Sect. 2.5) should be in the foreground. The relatively large amount of empirical data shows that a few untypical results have usually no major impact. Therefore it is expected that the overall results reflect systematic patterns, which can be theoretically explained. Tools to identify such patterns might be "exploratory data analysis" (from statistics) and empirical generalizations (see Sect. 4.3.4). "Exploratory data analysis is a descriptive pattern-detection process that is a precursor to the inductive generalizations involved in phenomena detection." (Haig 2013, p. 10).

One popular example for a generalization in marketing is the so called "experience curve", which implies that the unit costs for a product are assumed to decrease in line with increasing experience in manufacturing, logistics, and marketing of a product. This relationship was explored on the basis of a number of empirical studies by the Boston Consulting Group (see e.g. Tomczak et al. 2018).

Hubbard and Lindsay (2013, p. 1380) explain a central idea of theorizing on the basis of empirical generalizations:

"Successful theoretical interpretation typically comes after a pattern (fact) has been empirically determined. The rationale for this is that explaining particular or solitary events (e.g., individual decision-making) is likely to be unsuccessful because the events tend to be affected by idiosyncratic boundary conditions that are extremely difficult to establish. A better strategy is to anchor theory development around the detection of repeatable facts or regularities in the behavior of phenomena; their relative durability invites an explanation."

- **"Grounded Theory"**

A third way of forming theories is influenced, on the one hand, by experiences from marketing research practice and, on the other hand, by research strategies in other social science disciplines. In marketing research, it has long been common practice to address a novel problem with qualitative (pre-) studies. For example, focus groups, case studies, depth interviews, etc. serve to substantiate the research objectives and to prepare the methodology for a larger main study (e.g., Iacobucci and Churchill 2010). Similar in method, but with a different orientation and a different philosophy of science background, is the approach of the so-called

Table 4.2 Approaches of theory building and roles of empirical data

	Approaches of theory building		
Characteristics	Theoretical-in-isolation	Empirical-then-theoretical	Grounded theory
Use of empirical data	(Only later for theory test)	Basis for explanations and theory building	Interaction of theory building and empirical data
Type of data used	–	Quantitative	Qualitative
Amount of data (number of cases)	–	Large	Small

"**grounded theory**". The term indicates that in this approach a theory is "grounded" on the extraction and interpretation of empirical observations. This usually involves qualitative research methods. Data collection and theory-oriented interpretation of the observations are closely integrated and mutually influential (for details see Sect. 4.3.3). Table 4.2 summarizes the roles of empirical results in the three approaches of theory building discussed here.

The focus of this section is about theory building; the test of theories is discussed in the following chapter. In addition to theory formation and theory testing, the *modification* of theories is also relevant to research practice (see Sect. 9.4).

4.3.2 A Model of the Development of Theories as a Creative Act

The model of theory building outlined here refers to the "theoretical-in-isolation" approach and builds on the "inductive realist model of theory generation" developed by Shelby Hunt (2013, 2015). This model combines the presentation of processes of theory building and theory testing; Chap. 5 will deal with the latter. Therefore, we now focus on the part of the model that relates to theory building. We have made some modifications to the model designed by Hunt (2013). Figure 4.3 shows the model that we explain in this section.

First of all, we explain the "boxes" (1–8) in the model depicted in Fig. 4.3 according to Hunt (2013, 2015):

1. **Current disciplinary knowledge**: This box represents the current state of knowledge of a discipline (e.g., management research). This includes "*entities*" (e.g., companies, managers, customers) for which we commonly use theoretical concepts (see Sect. 4.1). These items have relevant "*attributes*" in each context, such as the size of the companies, the professional experience of the managers, or the frequency with which customers order. In addition, "*relationships*" exist between the entities, for example, large companies often have more managers or more specialized managers than smaller companies. Certain types of relationships become laws or lawlike generalizations (see Sect. 2.3.1) and certain relationship structures become theories (Hunt 2013). In addition, there are certain research traditions and methodical focuses in a discipline. For instance, a

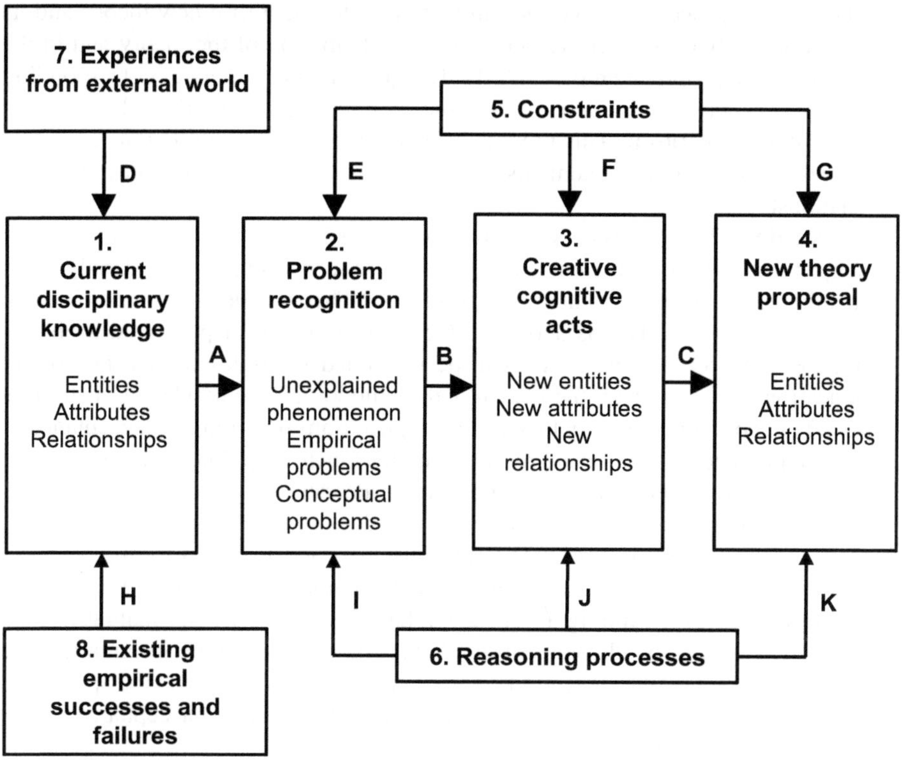

Fig. 4.3 Model of theory generation (adapted from Hunt 2013, p. 64)

behavioral orientation exists in management and marketing research; in the field of accounting and taxation, of course, the relevance of law is particularly great. Associated with this there are also certain methodological emphases in a discipline, in empirical marketing research, for example, the predominantly quantitative orientation.

2. **Problem recognition**: The identification of new and relevant research questions and the answers to these questions are at the core of scientific activities. This may be related to a hitherto unexplained phenomenon (e.g., effects of Internet use on the price sensitivity of consumers), to the lack of empirical confirmation of previously accepted theories, or to a conceptual problem (e.g., logical inconsistency of an existing theory or contradictions between two previously accepted theories).
3. **Creative cognitive acts**: This does not imply that theory generation is usually founded only on a sudden (more or less ingenious) inspiration. Rather, one turns towards a (time-consuming) process in which researchers develop new concepts (e.g., "electronic word of mouth"), observe previously unobserved properties (e.g., credibility of information sources on the Internet) or analyze new relationships (e.g., effects of corporate social responsibility on corporate goals).

The creative process involves not only the development of a new theory and its components, but also creative acts in the substantiation of the theory and in the creation of appropriate empirical tests. The quantity and variety of corresponding ideas have a positive influence on the theory building process (Weick 1989).

4. **New theory proposal**: This box represents the results of the previous cognitive processes. It contains statements about entities as well as their attributes and relationships.
5. **Constraints**: The process of problem recognition and theory building is typically subject to certain constraints. Some of these constraints have already been mentioned in Sect. 3.2 under the headings "Theory-ladenness" and "Social/ historical context". This is about the fact that the range of perceived problems and new theoretical approaches can be restrained through experiences, former education of researchers, theoretical and methodological knowledge or through social or economic pressure. In addition, expectations regarding the acceptance of new approaches in the academic community (such as publications and career opportunities) may also have constraining influences.
6. **Reasoning processes**: In science, creativity does not take place—as it does in some artistic areas—in total freedom; rather it is accompanied by the development of comprehensible and well-founded arguments. Therefore, the creative process of theory building is closely interlinked with the substantiation and evaluation of specific elements of the theory. At the least in the formulation and publication of new theories, a substantiation of their statements is indispensable, because otherwise no publication is possible and there is no acceptance by the academic community.
7. **Experiences from external world**: Experiences in reality show which phenomena have not been sufficiently researched and require appropriate theorizing.
8. **Existing empirical successes and failures**: The extent to which the current state of knowledge has proven its worth in empirical investigations (see Sect. 5.3) significantly influences the acceptance of the current state of knowledge of a subject area. Lack of success tends to lead to problem recognition and the goal of new theory building.

Below are brief explanations of the connections (A–K) between the different elements of the model:

- **A, B, C**: Here is the (ideal-typical) sequence of steps of theory generation. This is a simplified model (Hunt 2015) that does not include feedback processes.
- **E, F, G**: The "constraints" discussed above relate to problem recognition (e.g., critical evaluation of marketing practices), creative cognitive acts (e.g., influence of theory-ladenness), and the new theory proposal (e.g., limiting its degree of complexity).
- **I, J, K**: Accordingly, "reasoning processes" are required for problem recognition (e.g., relevance of the research question), creative cognitive acts (e.g., for assumed relationships), and—not least—for a new theory proposal (e.g., references from the literature).

- **D, H**: Here, the influence of experiences from the external world and the extent of the previous empirical successes and failures on the assessment of the current state of knowledge are present.

Chapter 5 (Sect. 5.3) introduces Shelby Hunt's "inductive realist model of theory status", which is closely related to his model of theory generation. This section presents only a part of this model. For a more comprehensive discussion, please refer to the corresponding articles by Hunt (2013, 2015).

4.3.3 Using Grounded Theory for Theory Building

The discussion about the advantages and disadvantages of so-called quantitative (e.g., representative surveys, experiments) and qualitative methods has been conducted intensively—sometimes fiercely—in the social sciences for years. These two approaches are fundamentally different in aspects of philosophy of science and research strategy views (see, for example, Hunt 2010; Neuman 2011). If one assumes—as in this book—a position of scientific realism, then the focus of *qualitative* research is clearly in the development of theories, while theory testing mostly applies to the so-called *quantitative* methods. Even in application-oriented studies, it is assumed that often in the first phases of the study an understanding of the problem has to be developed, for which qualitative methods are more appropriate, because most quantitative methods require a certain degree of understanding of the problem (including appropriate theoretical considerations), for example, for the research design and the development of measures.

Philosopher Gerhard Schurz (2014, p. 37) comments on the dispute over qualitative vs. quantitative methods:

"The ideological polarization between quantitative and qualitative methods that is held by some qualitative researchers (. . .) appears unnecessary and exaggerated. Rather, qualitative and quantitative methods are complementary. The strength of qualitative methods (e.g., case studies, narrative interviews) lies in advance of quantitative methods—in the exploration of relevant parameters and the generation of promising hypotheses. But a qualitative exploration has to be followed up by a quantitative-statistical analysis, as this is the only reliable way to test the generality of one's hypothesis, especially in a situation in which one does not already possess pre-established background knowledge. That qualitative and quantitative methods are complementary in the explained sense is a widely held view among empirical researchers in the social sciences (. . .); however, this view is not uncontroversial (. . .)."

Concerning the interplay between empirical data and theory formation, the **grounded theory** approach has attained special prominence (see, for example, Jaccard and Jacoby 2010, pp. 256ff.). The term "grounded" refers to the fact that in this approach theory does not arise only through more or less abstract considerations, but is developed on the basis of empirical observations. This approach goes back to Glaser and Strauss (1967). Corbin and Strauss (1990, p. 5) identify the central idea in the following way: "The procedures of grounded theory are designed to develop a well-integrated set of concepts that provide a thorough theoretical explanation of social phenomena under study. A grounded theory should explain as well as describe." Important and characteristic is the relationship between theory building and empirical data. "This approach emphasizes an approach of letting theory emerge from data rather than using data to test theory" (Jaccard and Jacoby 2010; p. 256).

The basic idea of the procedure for using grounded theory will probably be particularly clear in comparison to the (deductive) theory test (see Chap. 5). Figure 4.4 shows the fundamentally different goals and procedures of both approaches. In the *deductive* theory test, there is an already existing theory at the beginning, from which individual hypotheses are derived ("deduced") (see Chap. 5). These hypotheses predict to a certain extent the relationship between the variables involved in the external world (if the theory is true). Appropriate methods help measure these variables in reality; they are analyzed with statistical methods and the results allow the assumption of a confirmation or rejection of the hypothesis, which in turn corresponds to a "success" or a "failure" of the respective theory of interest.

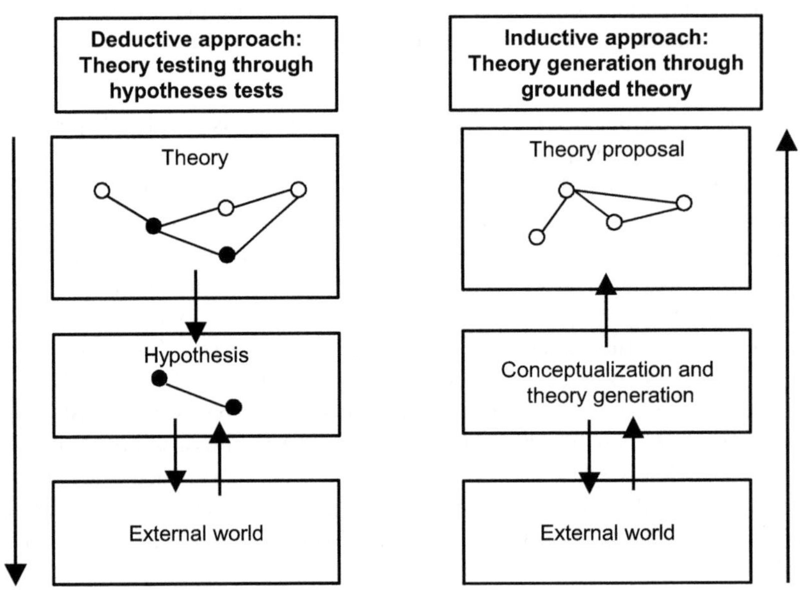

Fig. 4.4 Comparison of deductive theory test and inductive theory generation with grounded theory

In the case of *inductive approaches* of theory generation with grounded theory, the—as far as possible—unbiased (i.e., not theory-laden) observations of numerous aspects of a real phenomenon are the starting point. Based on this, concepts (see Sect. 4.1) for the relevant phenomena are developed. Assumptions about relationships between the various concepts then lead to building blocks of theories, which in turn are combined into a theory proposal. In Fig. 4.4 the arrows pointing in both directions between the fields "external world" and "conceptualization and theory generation" indicate that the latter should be in continuous feedback to the observations in reality (see above).

What is the methodological aspect most characteristic of grounded theory? There are various views in the literature, but there is broad consensus on essential principles (see below). Above all, with regard to the role of prior knowledge—especially from the literature—in the generation of theories, different views are present. Some authors believe that theory generation should be influenced by as little pre-information as possible in order to avoid "channeling" thinking (Jaccard and Jacoby 2010, p. 260) and to allow for openness to novel insights. On the other hand, scientists also suggest that a comprehensive literature knowledge of the interpretation of observations and their theoretical generalization is helpful. In this context, reference can be made to the problem of "theory-ladenness" in Sect. 3.2.

Jaccard and Jacoby (2010, p. 257) on the extent to which prior knowledge should be prominent in research:

"Early writings on grounded theory emphasized that researchers were to set aside, as much as possible, preconceived ideas that they have about the phenomenon of interest and instead let relevant concepts and relationships emerge from rich qualitative data. In later years some grounded theorists have maintained this orientation, whereas others have encouraged the use of prior knowledge and cognitive heuristics to help explore the nature of meanings (. . .)."

Let us now turn to the various methodological principles of grounded theory, of which the most prominent will be briefly presented here, based on Corbin and Strauss (1990). For the purpose of illustration, we add examples (brief corresponding quotes) from studies using grounded theory.

- *Data collection and data analysis are closely intertwined.* This is different from the typical procedure in other studies: "data collection → data analysis → interpretation". Rather, findings gained during data collection are analyzed immediately and will be used in the next steps of data collection (for example, in the next interviews). In this respect, the study design and its various details are typically not determined at the beginning of a study (Yin 2011, p. 77).

Lynn Isabella (1990, p. 13): "During the data collection phase at the organization studied here, notes on the facts, specific details, and other pieces of information that a number of participants seemed to repeat augmented the evolving theory (. . .), as did ideas generated during periodic debriefing sessions with colleagues."

- *Conceptualizations are the basic steps to theory generation.* Conceptualization also refers to the conceptual and abstracting summary of real phenomena (for example, behaviors or attributes) (see Sect. 4.1).

James Jaccard and Jacob Jacoby (2010, p. 271): "She then read each interview in earnest, placing a color-coded tag next to any segment that dealt with gender dynamics, and so on for each category of her typology."

- *Summary and linking of concepts to theoretical building blocks.* This process is the second stage of the process of abstraction of concrete perceptions. This concerns summaries and designations of previously developed concepts and considerations about a network of relationships of influencing factors and effects (Corbin and Strauss 1990).

John Holland (2005, p. 251): "The refined code networks were then used to suggest theoretical constructs and associated maps of causal elements that were constructed into a theory of corporate disclosure in the information market context (. . .)."

- *Selection of cases, informants etc. ("sampling") especially with regard to theoretical enrichment.* An (even approximately) representative sampling is not intended here. Rather, it is about "interesting" cases that bring new insights and also show the limits of these insights. The (targeted) selection of further objects of investigation takes place in the research process depending on the current state of knowledge according to criteria of the respective interests of the researchers ("theory-oriented sampling"). The data collection is terminated when additional objects of investigation promise no further increase in knowledge ("theoretical saturation").

John Holland (2005, p. 250): "Although this sample of companies provided a relatively high proportion of companies from the FTSE 100 (Financial Times Stock Exchange Index), the aim was not to provide 'statistical generalization' as in more conventional hypothetical-deductive research (. . .). The aim was to generate enough company cases to create the conditions for 'theoretical saturation' as recommended by Strauss and Corbin (. . .) (i.e., the point in category development at which no new properties, dimensions, or relationships emerge during analysis)."

- *Ongoing comparisons of research objects or of developed concepts.* Both concepts and cases should be compared with earlier developed concepts and cases studied so far in the research process with regard to similarities or differences. This should lead to a clarification of the conceptualization or the specific selection of further cases ("theoretical sampling"). In this sense, data collection and analysis are closely intertwined.

John Holland (2005, p. 251): "During the processing stages the interview responses of the various subjects were compared, continuously sampled, coded, and compared to each other, using the constant comparative method as recommended by Strauss and Corbin (. . .)."

- *Ongoing creation and archiving of notes ("memos") in the research process.* The developing thoughts on the research process, the development of concepts and steps in theory generation should be written down in a continuous and comprehensive manner in order to make the process and the reasons of theory generation comprehensible ("grounded"!).

John Holland (2005, p. 251): "These resulting codes were then checked to demonstrate that they were connected to original quotations in the source material and thus provided grounding. Codes such as 'private disclosure', the 'company story', or 'understanding state', or 'fragility' were therefore grounded in the original case data."

- *Coding is not considered a preliminary stage to data analysis, but is an integral part of data analysis.* In quantitative studies, the process of coding, that is, the translation of the information collected in appropriately selected symbols (usually numbers), is routine work and there exist certain exact rules that are applied as carefully as possible. By contrast, when using grounded theory, coding is a theoretically and methodologically demanding process that also requires creativity in abstraction and generalization based on a large and diverse set of individual pieces of information.

> Lynn Isabella (1990, p. 13): "I continually modified these initial categories, eliminating old ones and adding new ones to account for newly acquired evidence."

In an editorial for the Academy of Management Journal, Roy Suddaby (2006) has compiled some misunderstandings regarding grounded theory, which are presented here for further clarification:

- "*Grounded theory is not an excuse to ignore the literature.*" (p. 634). Apart from the question of whether it is even possible to liberate oneself from knowledge about and experience of prior literature, ignorance leads to less structured—and thus theoretically less fruitful—results with a low chance of publication. However, it is very important that pre-information does not limit the openness of the researcher.
- "*Grounded theory is not presentation of raw data.*" (p. 635). On the one hand, the results of a grounded theory application should be supported by collected data; on the other hand, grounded theory also includes *abstraction* in the formation of concepts or categories.
- "*Grounded theory is not theory testing, content analysis, or word counts.*" (p. 636). Neither the data collection nor the data analysis in the grounded theory approach would allow the testing of theoretical statements for their correspondence with reality. The scope of grounded theory lies rather in the more or less creative process of theory generation.
- "*Grounded theory is not simply routine application of formulaic technique to data.*" (p. 637). The central components of grounded theory are the interpretation of data and creative theory generation, both of which are processes that are certainly not standardizable and require a substantive understanding of the object of investigation.
- "*Grounded theory is not perfect.*" (p. 638). Grounded theory rules are not always clear and are not applicable in a schematic way, for example, in terms of theoretical saturation, that is, when the selection of additional cases can be finished.
- "*Grounded theory is not easy.*" (p. 639). The rather low formal requirements of grounded theory in comparison with some advanced statistical methods should not lead to the misapprehension that this is to be applied without much prior knowledge. Rather, appropriate experience, careful work and creativity are required.
- "*Grounded theory is not an excuse for the absence of a methodology.*" (p. 640). In the case of grounded theory (and other qualitative approaches), one sometimes finds the misconception that an "anything goes" rule applies. But the relatively high degree of methodological freedom requires careful documentation and justification of the methods used.

Marketing research that applies grounded theory uses these particular techniques for data collection:

- **Qualitative Interviews**: This refers to relatively long, unstandardized or only slightly standardized interviews, with which longer lines of thought or reasoning are collected and the respondents are encouraged to make appropriate reflections and to express them (see, for example, Yin 2011, pp. 134ff.).
- **Qualitative observations**: In doing so, the observer perceives attributes, behaviors and processes with his or her senses (especially, of course, visually and acoustically), without the need for verbal communication (see, for example, Yin 2011, pp. 143ff.). As a rule, the data collected are linked to the observation time or period.
- **Review of archived documents**: In particular, in organizations (e.g., companies, government agencies) there are extensive records in the form of correspondence, protocols, reports, etc. that can provide information about past events and processes.
- **Case studies**: Case studies may relate to processes (e.g., innovation processes), individuals (e.g., brand loyalty development), organizations (e.g., structure and strategy), or other social entities (e.g., families, informal groups). The subject of a case study are real phenomena, not artificially created or hypothetical ones. Typical for a case study is the use of different data sources and survey methods for a comprehensive and in-depth analysis of the case (Yin 2009; Morgan 2014).
- **Group discussions (Focus group interviews)**: This refers to the simultaneous questioning of several (often 6–10) respondents who are allowed to interact with each other. This corresponds to a more natural conversation situation and the participants stimulate each other.

Of course, the important aspect of the connection between empirical data and theory generation in grounded theory is particularly interesting from a methodological point of view. To a certain extent, it proceeds in an iterative manner and leads to a theory draft through a series of steps of theory generation and empirical observations. Figure 4.5 indicates that data collection and analysis intertwine closely in such a research process: At various points in the theory building process, researchers need to decide whether further data collection is helpful or necessary; newly collected data imply that the theory-building process must be continued or modified. The end of the process is a theory proposal, which can be tested later using the standard procedures of theory testing (see Chap. 5).

4.3.4 Empirical Generalizations and Theory Building

An *empirical generalization*, according to Bass (1995, p. G7), is: "a pattern or regularity that repeats over different circumstances and that can be described simply by mathematical, graphic, or symbolic methods. The definition does not assert causality and it does not require that the values of the parameters governing the

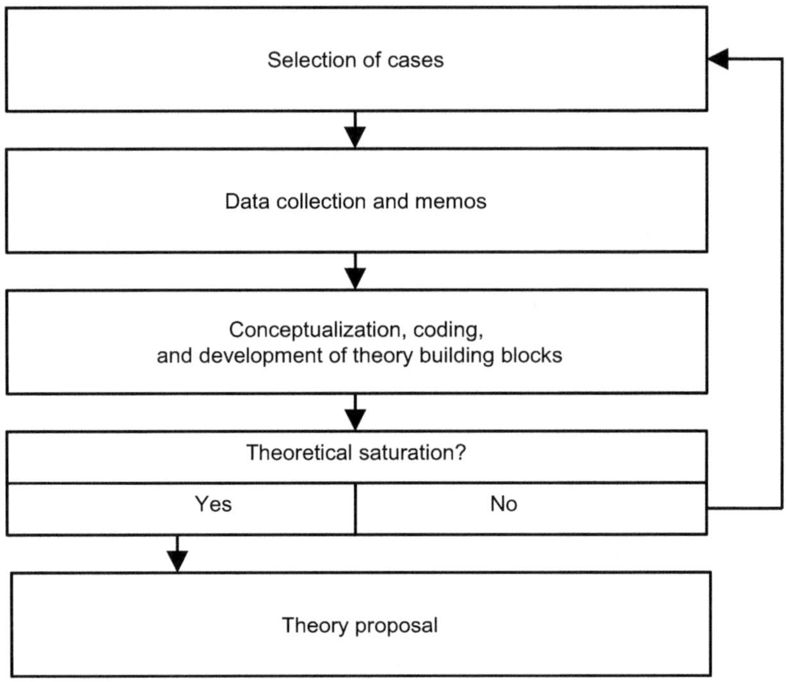

Fig. 4.5 Research process using grounded theory

regularity be invariant over the different circumstances. It does require that there be a pattern, but it does not require that the pattern be universal over all circumstances."

Here is an example of an empirical generalization and its implications. It is a meta-analysis (see Chap. 9) of a total of 114 studies on the impact of market orientation, whose results Dominique Hanssens (2009, p. 5) summarizes as follows:

"Market orientation (i.e., the organizational activities related to the generation and dissemination of and responsiveness to market intelligence, as well as the organizational norms and values that encourage behaviors consistent with market orientation) has a positive effect on organizational performance ($r = .32$), as measured by profits, sales, and market share. The market orientation–performance correlation is higher in manufacturing businesses ($r = .37$), compared to service businesses ($r = .26$). The association is stronger in countries that are low rather than high in power distance (i.e., how society deals with the fact that people are unequal in physical and intellectual capabilities) ($r = .33$ versus $r = .27$) and uncertainty avoidance (i.e., the extent to which a culture socializes its members into accepting ambiguous situations and tolerating uncertainty) ($r = .34$ versus $r = .27$)."

Because empirical generalizations can only generalize on the basis of existing data, they *do not claim universal validity*. On the other hand, empirical generalizations, of course, benefit from the broadening of the empirical basis and variety of studies on which empirical generalizations often rely, as well as by the multitude of researchers who perform these studies. This diversity serves as triangulation of empirical generalizations: different studies with different methods and data help to clarify how far the generalizability reaches (Kamakura et al. 2014).

Bass and Wind (1995, p. G2) summarize the following typical features of empirical generalizations:

"– Multiple studies: Minimum of two studies.
– Quality: The studies have to be of high quality.
– Objectivity: The studies should be by more than one author.
– Consistency: The results should be consistent under diverse conditions."

Empirical generalizations may also be useful without theoretical explanation. Isaac Newton's law of gravitation, which makes a statement about the effect of forces between two bodies, is an example of a very successful empirical generalization, which was without a theoretical justification for a long time, because it took more than two centuries before Albert Einstein, with his theory of relativity, theoretically explained gravitational interactions. When empirical generalizations are linked with theories, they can serve for both theory building and theory testing. *Theory building* attempts to theoretically explain or justify the empirical generalization determined by data, as in the example of the law of gravitation. In *theory testing*, empirical generalizations help to reduce the problems of testing hypotheses based on single studies. Results of empirical generalizations are less likely to suffer from the errors and limitations of "single-shot" studies.

The literature also discusses whether empirical generalization *can or should be relevant to marketing problems*. Precourt (2009, p. 113) explains the following points supporting the relevance of empirical generalizations for research and practice:

- Empirical generalizations serve as a *starting point for strategy development*. For example, the findings of the experience curve effect—a well-known empirical generalization—can be the starting point for the planning of the output quantity over time.
- Empirical generalizations provide preliminary *rules for management practice*. The experience curve effect offers a rule about the expected cost reduction over time.
- Empirical generalizations provide *benchmarks* for consequences of decisions or changes in planning. Empirical generalizations in the form of elasticities, for example advertising elasticities, provide an orientation for the expected sales changes with a change in the advertising budget.

- Empirical generalizations *serve as a guideline for future research*, as they show, for example, which results are to be expected with regard to a particular variable relationship.

Here is an example of the managerial implications of the results of the empirical generalization outlined in the above example (Hanssens 2009, p. 5):

"Market orientation provides a competitive advantage that leads to superior organizational performance. Even though the implementation of market orientation demands resources, it generates profits over and above the costs involved in its implementation, while concurrently growing revenues. This impact is greater in manufacturing businesses than in service industries. The implementation of market orientation processes should be adapted to local cultural sensitivities."

Empirical generalizations often become **laws** in the natural sciences, e.g., the already mentioned Newtonian law of gravitation. Social phenomena are usually more complex and dependent on a variety of influencing factors. Therefore, the social and behavioral sciences cannot fully explain repeated empirical observations simply by an underlying rule or formula, that is, a law. However, empirical generalizations may lead to **lawlike generalizations**. For this, the empirical data must be consistent with the expected values calculated on the basis of the underlying model or the underlying formula. In addition, empirical generalizations must provide not only a summary description of observations, but also a scientific explanation (see also Sect. 2.3.2).

References

Bass, F. M. (1995). Empirical generalizations and marketing science: A personal view. *Marketing Science, 14*(3), G6–G19.

Bass, F. M., & Wind, J. (1995). Introduction to the special issue: Empirical generalizations in marketing. *Marketing Science, 14*(3), G1–G5.

Corbin, J., & Strauss, A. (1990). Grounded theory research: Procedures, canons, and evaluative criteria. *Qualitative Sociology, 13*, 3–21.

Ehrenberg, A. S. C. (1993). Theory of well-based results: Which comes first? In G. Laurent, G. L. Lilien, & B. Pras (Eds.), *Research traditions in marketing* (pp. 79–108). Boston: Kluwer.

Glaser, B., & Strauss, A. (1967). *The discovery of grounded theory*. Chicago: Aldine.

Hacking, I. (1982). Experimentation and scientific realism. *Philosophical Topics, 13*, 71–87.

Haig, B. (2013). The philosophy of quantitative methods. In T. Little (Ed.), *The Oxford handbook of quantitative methods* (Vol. 1, pp. 7–31). Oxford: Oxford University Press.

Hanssens, D. M. (Ed.). (2009). *Empirical generalizations about marketing impact: What we have learned from academic research*. Cambridge, MA: Marketing Science Institute.

Holland, J. (2005). A grounded theory of corporate disclosure. *Accounting and Business Research, 35*, 249–267.

Hoyer, W. D., MacInnis, D. J., & Pieters, R. (2013). *Consumer behavior*. Boston: Cengage.

Hubbard, R., & Lindsay, R. (2013). From significant difference to significant sameness: Proposing a paradigm shift in business research. *Journal of Business Research, 66*, 1377–1388.

Hunt, S. D. (1987). Marketing research—Proximate purpose and ultimate value. In R. Belk, G. Zaltman, & R. Bagozzi (Eds.) (pp. 209–213). Chicago: American Marketing Association.

Hunt, S. D. (2010). *Marketing theory—Foundations, controversy, strategy, resource-advantage theory*. Armonk, NY: Sharpe.

Hunt, S. D. (2013). The inductive realist model of theory generation: Explaining the development of a theory of marketing ethics. *Academy of Marketing Science Review, 3*, 61–73.

Hunt, S. D. (2015). Explicating the inductive realist model of theory generation. *Academy of Marketing Science Review, 5*, 20–27.

Iacobucci, D., & Churchill, G. (2010). *Marketing research—Methodological foundations* (10th ed.). Mason, OH: South-Western Cengage Learning.

Isabella, L. A. (1990). Evolving interpretation as a change unfolds: How managers construe key organizational events. *Academy of Management Journal, 33*(1), 7–41.

Jaccard, J., & Jacoby, J. (2010). *Theory construction and model-building skills—A practical guide for social scientists*. New York: Guilford.

Jacoby, J., & Chestnut, R. (1978). *Brand loyalty—Measurement and management*. New York: Wiley.

Kamakura, W., Kopalle, P., & Lehmann, D. (2014). Empirical generalizations in retailing. *Journal of Retailing, 90*, 121–124.

Lilien, G., Kotler, P., & Moorthy, K. (1992). *Marketing models*. Englewood Cliffs, NJ: Prentice Hall.

MacInnis, D. J. (2011). A framework for conceptual contributions in marketing. *Journal of Marketing, 75*(4), 136–154.

MacKenzie, S. (2003). The dangers of poor construct conceptualization. *Journal of the Academy of Marketing Science, 31*(3), 323–326.

Magnani, L. (2009). Creative abduction and hypothesis withdrawal. In J. Meheus & T. Nickles (Eds.), *Models of discovery and creativity* (pp. 95–116). Dordrecht: Springer.

Morgan, M. (2014). Case studies. In N. Cartwright & E. Montuschi (Eds.), *Philosophy of social science* (pp. 288–307). Oxford: Oxford University Press.

Neuman, W. (2011). *Social research methods—Qualitative and quantitative approaches* (7th ed.). Boston: Pearson.

Nickles, T. (1985). Beyond divorce—Current status of the discovery debate. *Philosophy of Science, 52*, 177–206.

Nickles, T. (2000). Discovery. In W. Newton-Smith (Ed.), *A companion to the philosophy of science* (pp. 85–96). Oxford: Blackwell.

Nickles, T. (2008). Scientific discovery. In S. Psillos & M. Curd (Eds.), *The Routledge companion to philosophy of science* (pp. 442–451). London: Routledge.

Popper, K. (2002). *The logic of scientific discovery*. London: Routledge.

Precourt, G. (2009). Why empirical generalizations matter. *Journal of Advertising Research, 49*, 113–114.

Psillos, S. (2007). *Philosophy of science A–Z*. Edinburgh: Edinburgh University Press.

Schickore, J. (2014). Scientific discovery. In E. Zalta (Ed.), *The Stanford encyclopedia of philosophy*. http://plato.stanford.edu.

Schurz, G. (2014). *Philosophy of science. A unified approach*. New York: Routledge.

Suddaby, R. (2006). From the editors: What grounded theory is not. *Academy of Management Journal, 49*, 633–642.

Tomczak, T., Reinecke, S., & Kuß, A. (2018). *Strategic marketing. Market-oriented corporate and business unit planning*. Wiesbaden: Springer-Gabler.

Wacker, J. G. (2004). A theory of formal conceptual definitions: Developing theory-building measurement instruments. *Journal of Operations Management, 22*(6), 629–650.

Weick, K. (1989). Theory construction as disciplined imagination. *Academy of Management Review, 14*(4), 516–531.

Yadav, M. S. (2010). The decline of conceptual articles and implications for knowledge development. *Journal of Marketing, 74*(1), 1–19.

Yin, R. (2009). *Case study research—Design and methods* (4th ed.). Los Angeles: Sage.
Yin, R. (2011). *Qualitative research from start to finish*. New York: Guilford.
Zaltman, G., Pinson, C. R. A., & Angelmar, R. (1973). *Metatheory and consumer research*. New York: Holt, Rinehart and Winston.

Further Reading

Bass, F. M. (1995). Empirical generalizations and marketing science: A personal view. *Marketing Science, 14*(3), G6–G19.
Corbin, J., & Strauss, A. (2015). *Basics of qualitative research—Techniques and procedures for developing grounded theory* (4th ed.). Los Angeles: Sage.
Hunt, S. D. (2013). The inductive realist model of theory generation: Explaining the development of a theory of marketing ethics. *Academy of Marketing Science Review, 3*, 61–73.
Jaccard, J., & Jacoby, J. (2010). *Theory construction and model-building skills—A practical guide for social scientists*. New York: Guilford.
Schickore, J. (2014). Scientific discovery. In E. Zalta (Ed.), *The Stanford encyclopedia of philosophy*. http://plato.stanford.edu.
Yin, R. (2011). *Qualitative research from start to finish*. New York: Guilford.

5 Approaches for Theory Testing

5.1 Quality Criteria for Theories

In Sect. 3.1, the central criterion (at least from the perspective of scientific realism) for the assessment of a social science theory was addressed, namely its suitability for the description and explanation of real phenomena ("approximate truth"). Please refer also to a quote from Shelby Hunt (2010, p. 287) from Sect. 2.2: "When confronted with any theory, ask the basic question: Is the theory true?" This question is, of course, closely connected with the central function of empirical research, which is the subject of this book.

If one summarizes the results of empirical testing of a theory, then one speaks of the degree of *corroboration* of this theory. From this perspective, theories that have already been tested empirically several times (under different conditions) with a positive result are of higher quality than theories with a lower degree of corroboration. The **degree of corroboration of theories** plays an essential role with regard to the "status of a theory" in the inductive-realist model (see Sect. 5.3). It is primarily meta-analyses that allow a systematic assessment of the degree of corroboration of theories (see Chap. 9). An empirical test of theories is only possible if they are falsifiable. "It must be possible for an empirical scientific system to be refuted by experience" (Popper 2002, p. 18). The aspect of falsifiability has already been discussed in Sects. 1.1 and 3.1 as an essential criterion for the acceptability of statements. In essence, it is important that it must be *possible* for observations to contradict theoretical statements and refute the theory. Here are some simple examples of *non-falsifiable* statements:

- "Planning should be done carefully" (normative statement).
- "Brand loyalty occurs when at least 50% of all purchases of a customer in a product category are attributable to a single brand" (definition).
- "Even with increasing market share, profitability *may* decline" (immunized statement).

M. Eisend, A. Kuss, *Research Methodology in Marketing*,
https://doi.org/10.1007/978-3-030-10794-9_5

In addition to this fundamental requirement, there are more differentiated criteria for the evaluation of a theory. The basis for the following overview is the (more detailed) discussions in Jaccard and Jacoby (2010, pp. 31ff.), McMullin (2008), Sheth et al. (1988, pp. 29ff.) and Zaltman et al. (1973, pp. 91ff.).

- The first criterion is the **logical correctness** of a theory, its consistency. Any rational argument relies on the principle of consistency of statements. Logical consistency does not mean that a theory is true; but logical *inconsistency* would mean that the theory in question could not be true, because in contradictory statements at least one of them must be false.
- Furthermore, a high degree of **universality** of theoretical statements is desired. Universality does not refer to indetermination or lack of concreteness, but to the *scope* of a theory. The part of reality to which the statements of a theory refer should be as comprehensive as possible, not only in terms of spatial and temporal dimensions. In this sense, for example, a (general) theory of decision-making behavior would be considered "better" than a theory of decision making of marketing managers in industrialized countries. The universality of theories refers to the fundamental goal of science to make valid statements that go beyond a single individual case. For example, academic marketing researchers generally want to understand how to motivate salespeople, while practitioners are more interested in how a particular pay system affects the motivation of employees in their company (see also Sect. 2.5).
- The **precision** of a theory refers to a clear and unambiguous definition (see Sect. 4.1) of the concepts, as well as the equally clear and unambiguous formulation of the statements contained in a theory. This is by no means a trivial requirement. For instance, it was quite difficult to define the meaning of the involvement concept in consumer research (Zaichkowsky 1985). Jacoby and Chestnut (1978, pp. 57ff.) identified 53 (!) different brand loyalty measures at a relatively early stage of consumer research, many of which are based on different definitions of this concept. In such "chaos in researchland" (Jacoby and Chestnut 1978, p. 57), comparable, generalizable and unambiguously understood statements are hardly conceivable.
- Related to the idea of universality is the **information content** of a theory. This is high if the conditions ("if") for the occurrence of a phenomenon are very broad (these conditions occur relatively frequently) and the expectations based on the theory for the corresponding manifestations of this phenomenon ("then") are relatively concrete and accurate. Conversely, low information content refers to very specific conditions resulting in rather vague statements about the phenomenon of interest.

> An example of relatively high information content is:
>
> "If a company is one of the first five providers to enter a new market, its market share will be at least 10% after 3 years."
>
> An example of relatively low information content is:
>
> "If a company enters a market as a pioneer and has its technical know-how secured by patents, then it will still be present in the market after 3 years with a probability of $p > 0.1$."

- Furthermore, theories also require "**parsimony**", or simplicity. This means that they should include as few concepts, assumptions and statements about relationships as possible. Too much complexity would limit the comprehensibility and applicability of theories (Hunt 2015). Psillos (1995, p. 12) explains: "Simplicity is understood as minimizing the number of independently accepted hypotheses."
- The criterion of **originality** applies when a theory leads to completely new statements and thus greatly expands the existing scientific knowledge. A historical example of this is the theory of Nicolaus Copernicus (1473–1543) that the sun, rather than the earth, is at the center of our solar system. At this time, this theory decisively changed the worldview of humanity. In marketing research, new theories usually have far fewer revolutionary and less far-reaching effects. For instance, the Elaboration Likelihood Model (see, for example, Petty et al. 1983) is an example of originality, because it brought a novel and more comprehensive view of communication effects.
- **Fertility** is the ability of a theory to suggest ways to explore new phenomena and their relationships. For example, the Resource-Advantage Theory (see Hunt 2000, 2015) offers a comprehensive approach to understanding how to achieve competitive advantage, thus fertilizing many areas of marketing research.

5.2 Empirical Research to Test Theories

The procedure established in marketing research for the empirical testing of theories, that is, for assessing the correspondence between theory and reality, is the so-called **hypothetico-deductive method**. The basic idea is quite simple: From a general theoretical statement, empirical consequences for *concrete cases* are derived ("deduced") and these *expected* consequences are confronted with real observations. For example, from a general theoretical statement, "an increasing advertising budget leads to an increase in market share", one could conclude that at a certain time in a certain region an increase in the advertising budget would have the consequence that the market share increases (→ hypothesis). Depending on the consistency of the empirical observation with this hypothesis, one considers the theory to be supported or not.

Figure 5.1 illustrates the procedure for the hypothetico-deductive method. If, for example, a theory assumes a relationship between the concepts "A" and "B", then this would also have to be the case if one considers the relationship between specific expressions of A and B in reality (here "a" and "b"). From the theoretical conjecture ($A \rightarrow B$) a corresponding hypothesis ($a \rightarrow b$) was derived. As a rule, the validity of a hypothesis in reality is not accepted or rejected by mere inspection. In marketing research, this typically requires data collection and analysis methods that are applied during the *empirical research process*. This process will be discussed later in this section. Figure 5.1 shows, further, that after comparing the results that were expected by the hypothesis with the real observations, the hypothesis is accepted or rejected and the corresponding theory is more likely to be supported or questioned.

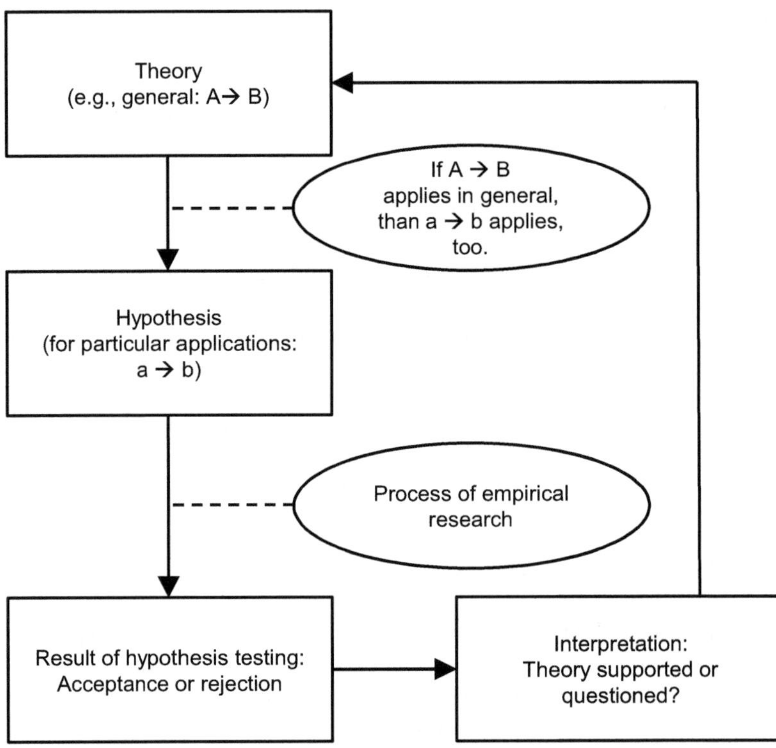

Fig. 5.1 The hypothetico-deductive method

Peter Godfrey-Smith (2003, p. 69) explains the basic ideas of the hypothetico-deductive method:

"We are dealing with a method rather than a theory of confirmation. Science textbooks are more cautious about laying out recipes for science than they used to be, but descriptions of the hypthetico-deductive method are still fairly common. Formulations of the method vary, but some are a combination of Popper's view of testing and a less skeptical view about confirmation. In these versions, the hypothetico-deductive method is a process in which scientists come up with conjectures and then deduce observational predictions from those conjectures. If the predictions come out as the theory says, then the theory is supported. If the predictions do not come out as the theory says, the theory is not supported and should be rejected."

The hypothetico-deductive method shown in Fig. 5.1 corresponds to Popper's (2002, p. 9) requirements for the test of theories by falsification attempts. However, both approaches have slightly different directions. The hypothetico-deductive method is fundamentally "neutral", but in research practice it is more likely to focus on empirical confirmation than rejection of hypotheses (or the theory behind them), while Popper's approach focuses on falsification attempts of existing theories and does not include their (inductive) confirmation. When describing the inductive-realist model in more detail (Sect. 5.3), it will become clear that affirmative and falsifying results are analyzed together and that, depending on the predominance of one or other kind of result, the acceptance of a theory is supported or not.

Karl Popper (2002, pp. 9–10) explains his view concerning deductive theory tests:

"According to the view that will be put forward here, the method of critically testing theories, and selecting them according to the results of tests, always proceeds on the following lines. From a new idea, put up tentatively, and not yet justified in any way—an anticipation, a hypothesis, a theoretical system, or what you will—conclusions are drawn by means of logical deduction. These conclusions are then compared with one another and with other relevant statements, to find what logical relations (such as equivalence, derivability, compatibility, or incompatibility) exist between them. (. . .).

Next, we seek a decision as regards these (and other) derived statements by comparing them with the results of practical applications and experiments. If this decision is positive, that is, if the singular conclusions turn out to be acceptable, or verified, then the theory has, for the time being, passed its test: we have found no reason to discard it. Nevertheless, if the decision is negative, or in other words, if the conclusions were falsified, then their falsification also falsifies the theory from which they were logically deduced."

Numerous empirical articles in leading marketing journals show that the hypothetico-deductive method is widely used in marketing research. However, major objections or restrictions regarding the informative value of this procedure are often overlooked. Three relevant aspects are:

- The first problem builds on the Duhem thesis, which was explained in Sect. 3.2. According to this, predictions of a theory are always bound by assumptions about the suitability of the observations and measurements made to represent the theoretically interesting phenomena. If a hypothesis is rejected, it may be because it is actually wrong, just as the opposite applies if it is correct *or* because erroneous observations/measurements have led to a result that does not reflect reality. Hypothesis tests are meaningful only if the observations or measurements actually refer to the theoretically interesting concepts, that is, if they are reliable and valid. This may sound trivial, but it represents a central problem in empirical research practice, which will be discussed in detail in Chap. 6.
- Typically, results of empirical studies are based on statistical conclusions (i.e., inferential statistics based on random samples). As a result, these results are subject to uncertainty, e.g., in the form of confidence intervals or error probabilities. When accepting or rejecting a hypothesis, such mistakes can occur. In many cases decisions about hypotheses are based on significance tests, which have specific problems and limitations (see Chap. 7).
- The third (often hidden) problem is that different theories could lead to the same predictions. "If the only source of empirical confirmation is by way of the verification of such predictions, then it is difficult to avoid the conclusion that all theories which entail the same predictive consequence receive exactly the same degree of confirmation from the prediction." (Sankey 2008, p. 252). Therefore the hypothetico-deductive method doesn't "discriminate between mutually incompatible but empirically equivalent hypotheses" (Psillos 2007, p. 114).

Characteristic of empirical research for theory testing is thus the comparison of theoretical assumptions with real observations. Figure 5.1 mentions the "*process of empirical research*". What is meant by this process? A model of the empirical research process presented below provides a conceptual framework and explains essential parts of the research process and their relationships. Several considerations from Chap. 2 and from Sect. 4.1 are taken up and summarized with regard to theory testing.

A basic **model of empirical research** is presented according to Kuss and Eisend (2010, pp. 18ff.). It is typical for the scientific view of reality that attempts are made to establish consistent systems of statements. Those statements, whose correspondence to reality should be established and checked, are referred to as theory if they meet certain conditions (see Chap. 2). Since these systems of statements usually have a high degree of complexity and/or abstraction that does not permit direct examination (for example, by simple direct observation), suitable empirical methods are used for this purpose. For example, the investigation of a correlation between risk perception and the need for information usually requires a rather elaborated research design (including measurement instruments, random sampling, statistical methods, etc.). By mere inspection alone, it is not possible to make such an assessment.

Next, we present the three basic elements of empirical research—reality, theory, and methods—followed by a discussion of their relationships.

- **Reality**

Regardless of the respective research interest, it is only possible to look at corresponding *parts* of reality. A complete description or explanation of reality is impossible, due to its particular properties. According to Jaccard and Jacoby (2010, p. 9f.) reality is complex, dynamic, (partially) obscured, and unique. These details are presented in detail in Sect. 2.2. The combination of these properties leads to the fact that empirical research can only refer to a few, selected parts of reality that are abstracted from some elements of an overwhelmingly complex reality.

- **Concepts, theories and hypotheses**

Because it is not possible to fully grasp reality (see also Sect. 2.2), the idea of empirical research is different. One makes use of certain abstractions of single phenomena, which are important for the respective point of view and that have already been referred to and discussed as "*concepts*" in Sects. 2.1 and 4.1. By simplifying and organizing the environment through concepts, one can discover certain regularities and relationships. These phenomena can be very concrete ("the larger a trade fair stand, the greater the number of visitors"), but also more abstract ("for technically trained trade fair visitors, economic criteria play a lesser role in the purchasing decision than for visitors with a business training "). Of course, systems of statements that include a larger number of concepts, and/or relationships between these concepts, are particularly powerful. These systems are *theories*, as described in the second chapter. Each theory uses several concepts (in the above example, these are "technical education" and "importance of economic criteria"). In this respect concepts form the "building blocks" of theories (Jaccard and Jacoby 2010, p. 11).

In connection with the testing of theories (or parts of theories), but also with practical questions, *hypotheses* play an important role—as has already been shown in the above presentation of the hypothetico-deductive method (see also Chap. 7). These are (not yet verified) assumptions about:

- Values of variables (e.g., "At least 10% of consumers will try the new product X" or "At least 80% of all companies with more than 5000 employees have a separate marketing department").
- Relationships of variables (e.g., "Superior resource facilities lead to above-average profitability of companies" or "The more positive the attitude to a product, the greater the propensity to buy").

The first type of hypotheses (values of variables) plays the larger role in applied research (e.g., market research), while the second type (relationships of variables) is more important for theory-related research.

Jaccard and Jacoby (2010, pp. 76–77) characterize hypotheses by three features. Hypotheses are

1. derived from a theory,
2. more concrete than the statements of the theory, and
3. oriented towards empirical research in such a way that they allow an empirical testing of theoretical statements.

In this perspective, hypotheses are links between theory and empirical data and methods, because they determine the relationships between the more abstract theory and the more concrete field of investigation. The formulation of hypotheses is also important with regard to the selection of methods appropriate for the research questions. If one thinks of one of the examples of a hypothesis outlined above, then one notices that it can be deduced directly which variables (for example, "attitude towards a product", "propensity to buy") must be measured and which measurement levels (see Sect. 6.2) are needed. This requires appropriate methods, the definition of which is the subject of the next step in the research process (Fig. 5.2).

- **Methods**

If theories or parts of theories are tested for their resemblance with reality, appropriate methods are needed (see Chaps. 6, 7, 8 and 9). In particular, theories with concepts of a high degree of abstraction often face difficult measurement problems.

It is therefore necessary to establish a connection between the (abstract) elements of theories and reality. Data collection methods in empirical research can also be seen as a tool to observe the interesting aspects of reality, despite their complexity

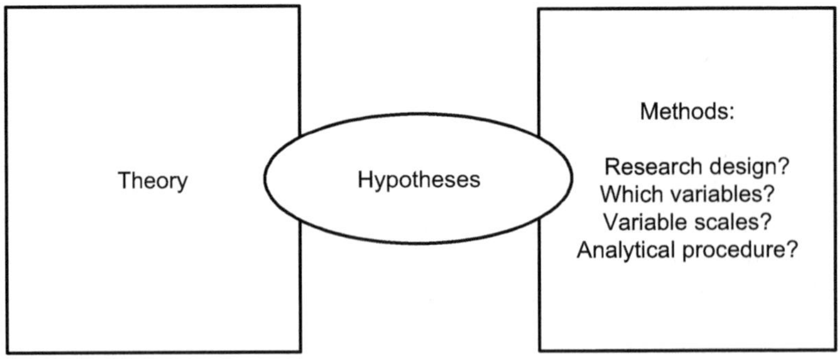

Fig. 5.2 Hypotheses as link between theory and methods

and (at least partial) obscurity. For example, the methods of sampling indicate which sub-set of objects to examine. As another example, questionnaires can be used to summarize different people, opinions, or behaviors into groups or categories (for example, people with a high level of education, or negative attitudes to television advertising) or classify them along corresponding measurement scales. The main purpose of data analysis procedures is to aggregate a large amount of individual data (e.g., by measures such as means or variances) and to reveal (statistical) relationships between variables.

The three elements of empirical research are presented in Fig. 5.3. The connecting arrows indicate fundamental sub-tasks in the *process* of empirical research, which will be discussed below.

Conceptualization is the process of identifying parts of reality that are of interest in an abstract manner (see Sect. 4.1). Often this process goes hand in hand with the development of assumptions about the relationships between these concepts and the formation of theories in the sense of an inductive approach. Conceptualizations lead to corresponding *definitions*, which precisely formulate what constitutes the respective phenomenon. On the one hand, therefore, there is an abstraction from reality; on the other hand, this abstraction also determines the *way of looking* at reality; therefore, the corresponding arrow in Fig. 5.3 points in both directions. For example, if one abstracts certain characteristics of the behavior of customers by the concept of "satisfaction", this concept also influences the perspective when looking at the real behavior of these customers.

In order to compare theories with reality, suitable methods need to be selected. For example, one has to decide with which scale to measure satisfaction, which might be the cause of customer loyalty. A statistical procedure also has to be selected to analyze the assumed relationship. This procedure is called **operationalization** (see Sect. 6.1). Concrete measurement methods, statistical procedures, etc. are thus assigned to abstract concepts. As a rule, this also involves narrowing down quite general concepts to concrete objects of investigation. Thus, one can hardly empirically examine the *general* relationship between customer satisfaction and customer

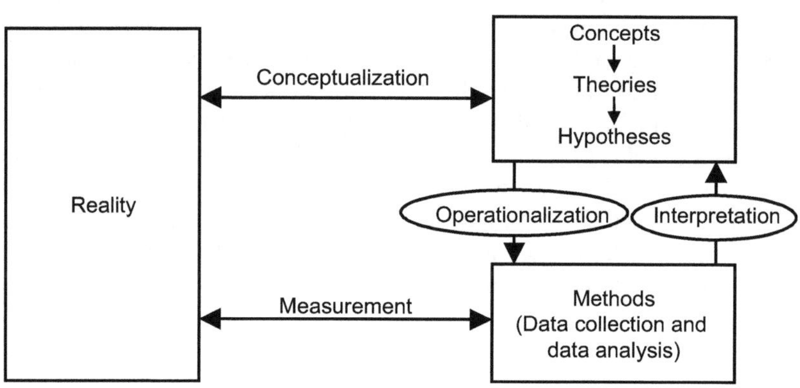

Fig. 5.3 Conceptual model of empirical research (Source: Kuss and Eisend 2010, p. 23)

loyalty. The focus must be on much more concrete and feasible—and thus less general—relationships (for example, the relationship between "customer satisfaction with product X" → "customer loyalty concerning product X").

Applying the selected methods of data collection to corresponding parts of reality is called **measurement** (see also Sect. 6.2). This process is also two-sided: subjects, objects, etc. are exposed to measurement instruments; data flow back. Nunnally and Bernstein (1994, p. 3) define: "Measurement consists of rules for assigning symbols to objects so as to (1) represent quantities of attributes numerically (scaling) or (2) define whether the objects fall in the same or different categories with respect to a given attribute (classification)."

These data can be summarized and displayed by using statistical methods (data analysis). The considerations, when comparing results of the data analysis with the statements of the theory, are called **interpretation**. It determines whether the theory or parts of it have been *confirmed or not* and whether *modifications* to the theory (see Sect. 9.4) need to follow. Here, the relationship to the "inductive-realistic" model of theory testing and the idea of "empirical successes and failures" (Sect. 5.3) is quite obvious. The *refinement* of a theory might be another outcome of the interpretation, for instance with the specification of relationships (linear or nonlinear) or the determination of elasticities (see Sect. 9.4).

For the basic model of empirical research presented here, the *requirement* is that test results that should answer a question or test a hypothesis can only be meaningful if the data collection and data analysis (including sampling, measurements, data processing, etc.) actually reflect the phenomenona under investigation. Notably (as explained in Sect. 3.2), measurement errors can significantly affect the results of theory tests, because it is not clear whether a lack of correspondence between a theoretical assumption and an empirical result relates to the measurement errors *or* is due to the defectiveness of the theoretical assumption (see above).

The requirement that the real phenomena considered in an empirical investigation should correspond as far as possible to the theoretically interesting concepts may seem trivial at first sight. In social science measurements, however, this problem is anything but trivial. Consider the following example of corresponding problems with measurements of the concept "purchasing behavior": If a consumer says that he or she wants to buy a brand, can one conclude that he or she will actually (always, mostly, occasionally) buy that brand? Is it possible to conclude, from the verbal report of consumers that they purchased a brand during their last shopping trip, that they did *actually* purchase the brand or do we have to deal with memory gaps, adjustments to the expectations of an interviewer or deliberately voiced misstatements that can all lead to measurement errors?

The question of whether the implementation of a problem in a research design (with sampling, measurement methods, etc.) is appropriate, is of the greatest importance for the informative value of empirical studies. Two basic problems can occur:

- Does the study, with all its methodological details, lead to systematic deviations from the "true values" of the phenomena to be examined? Example: Does the measurement of the environmental awareness of a population lead to a systematic overestimation through an appropriate survey, because many people (for example because of the social desirability of ecological awareness) tend to give too favorable answers to this question?
- Is the result influenced by coincidences (and negligence) during the research process? Example: Is it possible that the interview time (morning or evening, weekday or weekend) can lead to different information from respondents regarding their preferences for certain food, drinks or leisure activities?

This brings us to two basic criteria for the quality of empirical research: **Validity**, which refers to (as far as possible absent or very small) *systematic* deviations of test results from reality, and **reliability**, which is about the independence of results from a one-time measurement process. With high reliability, i.e., with low *situational* (more or less random) influences, repeated or similar measurements would always lead to the same (or at least very similar) results (if the values of the concept to be measured do not change). These important points follow in detail in Chap. 6.

With the aid of the basic model of empirical research, the application of the hypothetico-deductive method can now become more concrete. The process begins with the theory to be tested and the hypotheses derived from it. For their empirical testing, suitable data collection methods are developed by means of operationalization. These methods are applied to reality during measurements and one receives data that can be statistically analyzed. The results are interpreted in terms of confirmation ("success") or rejection ("failure") of the tested theory.

5.3 Inductive-Realist Model of Theory Testing

Hunt (2011, 2012) has summarized the procedure and conclusions of (empirical) research from the perspective of scientific realism within a model outlined below. It turns out that following the basic ideas of scientific realism, for example, the "no miracle argument" (see Sect. 3.1), that the consideration of theories (empirical successes and failures) is of central importance. First, a brief characterization of the model follows, then the model is presented in Fig. 5.4 and afterwards the corresponding explanations follow.

The term "inductive-realist model" already gives some information about its central features (Hunt 2015). What do the components of this term mean?

"Inductive" This term refers to the reasoning behind the model and can be easily understood through a quote by Ernan McMullin (1984, p. 26): "The basic claim

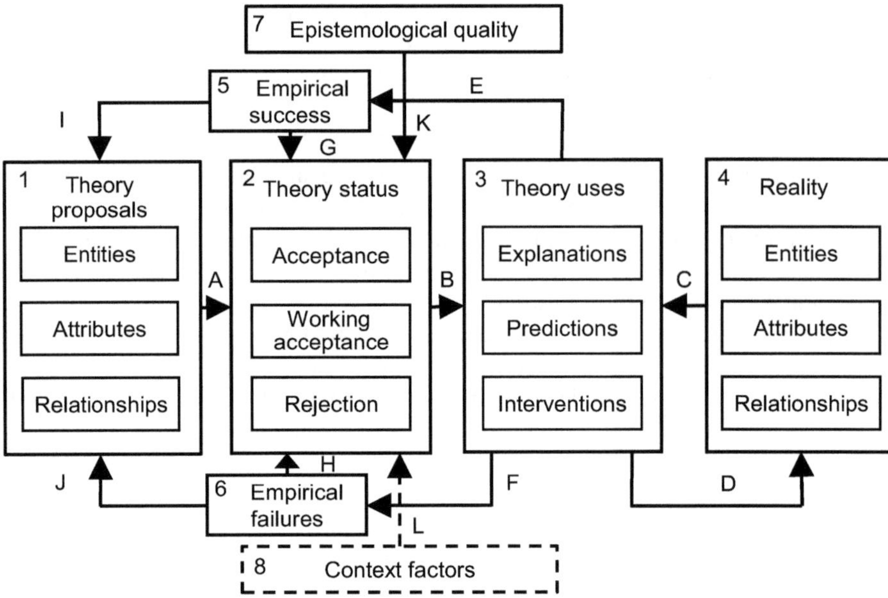

Fig. 5.4 Inductive-realist model of theory testing (Source: Hunt 2012, p. 9, with minor modifications)

made by scientific realism (. . .) is that the long-term success of a scientific theory gives reason to believe that something like the entities and structure postulated by the theory actually exists" (see Sect. 3.1). Here one clearly recognizes the features of inductive reasoning (see Sect. 2.5), because the "long-term success of a scientific theory" serves as the basis for the assumption of their (far-reaching) correctness. Furthermore, in the application of the model, empirical successes and failures are related to the acceptance or rejection of a theory.

"Realist" This term indicates that scientists make statements about a reality that exists independently of their perceptions and interpretations (see Sect. 3.1).

"Model" It is a model in the sense that the process of theory testing is depicted in a *simplified* form (see Sect. 2.1).

The explanations of Hunt's (2012) model are as follows: The four boxes 1 to 4 stand for the theory, the status of the theory, applications of the theory and reality (in the original "External World", which is used synonymously with reality, according to a personal information from Shelby Hunt). *Box 4* contains (real) "entities" (e.g., companies, brands), "attributes" (e.g., characteristics of the companies and brands), and "relationships" between entities, their attributes, and each other. Since theory and reality should correspond, it is not surprising that the content of Box 1 corresponds to that of Box 4. *Box 1* contains the concepts corresponding to the entities with their attributes in Box 4 ("Reality") and the theoretical conjectures about relationships. If theory and reality correspond sufficiently, the theory becomes (approximately) true (Hunt 2010, p. 287) (see Sect. 3.1).

Box 3 ("Theory uses") contains very different elements. These are the three principal applications of theories to real phenomena:

- Explanations (see Sect. 2.3.2). These are questions about the "why" of the occurrence of certain real phenomena. (Example: "Why are market pioneers often successful in the long run?")
- Predictions (including hypotheses). On the one hand, this involves the use of (theoretical) knowledge about "if-then relationships" for statements about future phenomena when certain conditions apply. (Example: "If customer satisfaction increases, then brand loyalty will increase too".) On the other hand, hypotheses are (theoretically based) assumptions about the characteristics and relationships of phenomena under certain conditions and, in this sense, are predictions for these conditions (see Sect. 5.2).
- Interventions refer to actions—often based on theoretical considerations or experiences—that influence or change reality. For example, by knowing the importance of online communication for young customers, a manager may decide to change the communication budget accordingly and may *change reality* in this way.

Box 2 ("Theory status") identifies different assessments of a theory in the respective "scientific community". The "acceptance" category means that a theory has been adequately tested and considered the best available theory for the particular area. Such a theory is most likely the basis for the explanations, predictions, and interventions mentioned in Box 3. The category "working acceptance" refers to theories that are not yet fully established and are still being developed and tested. Of course, corresponding hypothesis tests (see Chap. 7) are central to this process. It should not be surprising that "rejection" refers to theories which are not in use or only in a few exceptional cases.

Boxes 5 and 6 show the frequencies or proportions of successful and unsuccessful applications of a theory. Depending on the result, many "successes" lead to the strengthening of the assumption (Arrow G) that corresponding entities, attributes and relationships actually exist in reality *or* in the case of many "failures" to reinforce doubts regarding the truth of the theory (Arrow H).

Now to the other relationships contained in the model that are represented by arrows:

- Arrow A shows that, over time, theories are tested, more or less approved, and are ultimately accepted or rejected.
- Arrow B stands for the use of theories for explanation, prediction and intervention or for hypothesis tests of theories of "working acceptance".
- Arrow C symbolizes the "feedback" from reality on attempts of explanation, prediction or intervention and is set as success or failure.

- Arrow D shows that particular interventions (for example, a price reduction based on a prediction on the price-sales function) have a direct impact on reality (i.e., markets, competitors, etc.).
- Arrows E and F indicate that theory uses due to success (e.g., an accurate prediction → E) or failure (e.g., an intervention that does not have the expected result → F) will feed into the further assessment of this theory.
- Arrows G and H represent the effects of successes or failures on the increasing (G) or decreasing (H) acceptance of a theory.
- Arrows I and J stand for the effects of successes and failures on the further development of a theory through modifications, refinements, additions, etc.

In a recent version, Shelby Hunt (2012) has added two boxes and corresponding arrows to the inductive-realist model, which have nothing to do with the empirical test of a theory, but through which essential results of the philosophy of science discussions of the past decades are integrated into the model. Box 7, with the somewhat strange sounding name "epistemological quality" and Arrow K refer to the fact that the degree of acceptance of a theory depends not only on its empirical test, but also on other quality characteristics (e.g., logical correctness, precision, and information value), discussed in Sect. 5.1. Similarly astonishing at first glance is the term "context factors" for Box 8 with Arrow L. In this, Hunt takes up the arguments concerning the relevance of the social/historical context (see Sect. 3.2) or of theory-ladenness (see also Sect. 3.2). Through the dashed lines, he indicates that he sees in this not a contribution to truth, but rather the opposite. Further influencing factors in this sense are unethical behavior of scientists, for example data manipulation and negligence (see Chap. 10) as well as political or social norms and influences of funding organizations.

The well-known relationship between attitudes and buying behavior may illustrate the procedure and conclusions of the inductive-realist model (Hunt 2012). We assume that (the entities) "attitudes" and "buying intentions" exist with the attributes "negative/positive" and "weak or strong intent" respectively, as well as the corresponding relationship in reality (→ Box 4), that they are perceived by the researchers and lead to theory building (see Sect. 3.3). Now if such a theory exists (→ Box 1), then it can be used (→ Box 3) to find explanations, to develop predictions and to prepare interventions. Here are some examples:

- Different buying behaviors of different consumers are *explained* by varying attitudes.
- Based on a positive change of attitudes, a corresponding buying behavior is *predicted* or a corresponding *hypothesis* is tested in an empirical study.

(continued)

- A manager uses the theoretically assumed relationship between attitudes and buying behavior as the basis for an *intervention* (such as increased advertising) to change attitudes with the expected effect on buying behavior.

It shows then (Arrow C) whether the applications of the theory were successful:

- *Explanation*: Did the consumers who have bought a product actually hold attitudes that are more positive?
- *Prediction*: Has the share of buyers and shoppers actually increased according to the positive development of attitudes? Has the hypothesis been confirmed in the study?
- *Intervention*: Did the increase in communication lead to more positive attitudes and thus to increased sales?

Depending on the proportions of the "successes" and "failures" of the theory, the acceptance or tendency to modify or reject the theory increases or decreases (Box 2).

Essential for the theory status is thus the relation between "successes" and "failures" in theory testing. Theories with a clear preponderance of one or other kind of result (with a sufficiently large number of empirical tests), quite clearly lead to "*theory acceptance*" or "*theory rejection*". In situations where (too) few tests have been made or where there is no clear preponderance of "successes" or "failures", further empirical research would be the appropriate response ("*working acceptance*"). The inductive-realist model thus makes it clear that the process of increasing (or decreasing) empirical confirmation, from the perspective of scientific realism, is central to the degree of acceptance of a theory. For this reason, the approaches discussed in Chap. 9, related to the aggregation and integration of empirical results (meta-analyses, etc.), are relevant.

References

Godfrey-Smith, P. (2003). *Theory and reality—An introduction to the philosophy of science*. Chicago/London: University of Chicago Press.

Hunt, S. D. (2000). *A general theory of competition—Resources, competences, productivity, economic growth*. Thousands Oaks, CA: Sage.

Hunt, S. D. (2010). *Marketing theory—Foundations, controversy, strategy, resource-advantage theory*. London: Sharpe.

Hunt, S. D. (2011). Theory status, inductive realism, and approximate truth: No miracles, no charades. *International Studies in the Philosophy of Science, 25*, 159–178.

Hunt, S. D. (2012). Explaining empirically successful marketing theories: The inductive realist model, approximate truth, and market orientation. *Academy of Marketing Science Review, 2*, 5–18.

Hunt, S. D. (2015). Explicating the inductive realist model of theory generation. *Academy of Marketing Science Review, 5*, 20–27.
Jaccard, J., & Jacoby, J. (2010). *Theory construction and model-building skills—A practical guide for social scientists*. London: Guilford.
Jacoby, J., & Chestnut, R. (1978). *Brand loyalty—Measurement and management*. New York: Wiley.
Kuss, A., & Eisend, M. (2010). *Marktforschung. Grundlagen der Datenerhebung und Datenanalyse* (3rd ed.). Wiesbaden: Gabler.
McMullin, E. (1984). A case for scientific realism. In J. Leplin (Ed.), *Scientific realism* (pp. 8–40). Berkeley: University of California Press.
McMullin, E. (2008). The virtues of good theory. In S. Psillos & M. Curd (Eds.), *The Routledge companion to philosophy of science* (pp. 499–508). London: Routledge.
Nunnally, J. C., & Bernstein, I. (1994). *Psychometric theory* (3rd ed.). New York: McGraw-Hill.
Petty, R. E., Cacioppo, J. T., & Schumann, D. W. (1983). Central and peripheral routes to advertising effectiveness: The moderating role of involvement. *Journal of Consumer Research, 10*(September), 135–146.
Popper, K. (2002). *The logic of scientific discovery*. London: Routledge.
Psillos, S. (1995). Theory, science and realism. *Lecture Notes*. London School of Economics.
Psillos, S. (2007). *Philosophy of science A—Z*. Edinburgh: Edinburgh University Press.
Sankey, H. (2008). Scientific method. In S. Psillos & M. Curd (Eds.), *The Routledge companion to philosophy of science* (pp. 248–258). London: Routledge.
Sheth, J., Gardner, D., & Garrett, D. (1988). *Marketing theory—Evolution and evaluation*. New York: Wiley.
Zaichkowsky, J. L. (1985). Measuring the involvement construct. *Journal of Consumer Research, 12*(December), 341–352.
Zaltman, G., Pinson, C. R. A., & Angelmar, R. (1973). *Metatheory and consumer research*. New York: Holt, Rinehart and Winston.

Further Reading

Hunt, S. D. (2011). Theory status, inductive realism, and approximate truth: No miracles, no charades. *International Studies in the Philosophy of Science, 25*, 159–178.
Jaccard, J., & Jacoby, J. (2010). *Theory construction and model-building skills—A practical guide for social scientists*. London: Guilford.
Sankey, H. (2008). Scientific method. In S. Psillos & M. Curd (Eds.), *The Routledge companion to philosophy of science* (pp. 248–258). London: Routledge.

Obtaining Data for Theory Testing: Operationalization, Measurement, and Data Collection

6

6.1 Operationalization

If theories are to be tested for their correspondence with reality, appropriate methods are needed. It is often not easy to develop suitable methods for the empirical testing of theories and the measurement of concepts used in them, especially in the case of theories that involve concepts with a high degree of abstraction (e.g., dynamic capabilities of companies or managers, innovativeness of consumers). These concepts are usually not directly observable in everyday life and, thus, are not easily quantifiable (e.g., the extent of the dynamic capabilities of a company). As Sect. 5.2 already clarified, the main issue is to establish a connection between the (abstract) elements of theories and reality by means of empirical research methods. "Measurement is an activity that involves interaction with a concrete system with the aim of representing aspects of that system in abstract terms (e.g., in terms of classes, numbers, vectors, etc.)" (Tal 2015, p. 1).

During theory building, assumptions about relations of concepts are made. The process of **conceptualization** describes parts of reality that are of interest in an abstract way (see Sect. 4.1). A conceptualization leads to a definition that verbally expresses what constitutes the relevant phenomenon. In order to be able to confront these concepts with reality, one has to make them measurable. The process of making something measurable is called **operationalization** (Bridgman 1927). Operationalization determines how a theoretical concept should be observed or measured. Specific measuring methods are assigned to abstract concepts. For example, if one wants to measure a person's intelligence, so-called intelligence tests apply. The questions on an intelligence test represent the measurement of intelligence, and with the development of these tests, the abstract concept has been operationalized. Related to the process of operationalization is the narrowing of quite general concepts to concrete objects of investigation. For example, one can hardly empirically investigate the general relationship between satisfaction and behavior. Instead, researchers focus on much more specific—and, thus, less

M. Eisend, A. Kuss, *Research Methodology in Marketing*,
https://doi.org/10.1007/978-3-030-10794-9_6

general—relationships (for example, the relationship between "customer satisfaction" and "brand loyalty").

Operationalization thus assigns observable facts to theoretical concepts. Observable facts refer to variables. A *variable* is a characteristic or specific behavior of objects (e.g., the age of a person or the electoral behavior with regard to political parties), whose expression (e.g., 35 years old or election of a liberal party) can be unambiguously determined for different objects as a result of measurement. At the same time, the process of operationalization determines the variables that are considered in an empirical study.

One can distinguish between manifest and latent variables. **Manifest variables** can be directly observed or determined empirically because their characteristics can be determined by direct measurement of a relevant object. For example, the age of a person can be requested directly. By contrast, **latent variables** are not directly observable and cannot be measured directly, such as the intelligence of a person. Latent variables include all concepts with a certain degree of abstraction, such as the innovativeness of companies. But how can we operationalize and measure a latent variable? For this purpose, we use so-called indicators. **Indicators** are manifest variables that help to measure a latent variable by means of their operationalization. For instance, the latent variable "religiosity" is captured by indicators such as the frequency of prayers or church visits, which can be directly measured or requested. Obviously, we often use several indicators to fully capture a latent variable with its complexity and abstraction. It is easy to imagine that a concept such as intelligence cannot be meaningfully grasped with a single question (for example, a mere knowledge question about the birth year of a famous person) but that a multitude of questions are necessary. In the context of surveys, we often speak of **items** instead of indicators. In this book, we use "items" and "indicators" interchangeably.

Latent variables can also include **multiple dimensions**. A multidimensional latent variable or concept occurs when different, but related, dimensions are used to measure a concept (or construct). For example, the concept of the credibility of a person or information source includes two dimensions, namely, trustworthiness and competence. Several indicators operationalize and measure each dimension. Indicators such as "experienced," "professional" or "qualified" capture the competence dimension, and indicators such as "honest" or "open" outline the trustworthiness dimension. Multidimensional concepts can, therefore, be distinguished from unidimensional concepts. Figure 6.1 illustrates the difference between unidimensional and multidimensional constructs.

The "social class" is a classic example of a concept with several dimensions and different indicators for each dimension. It is usually captured by three dimensions: education, income, and occupation. The table below shows, for each dimension, two indicators and possible measurement instruments for the indicators.

Concept	Dimensions	Indicators	Instrument
Social class	Education	School education	"What is the highest level of school you attended?"
		Work education	"Which vocational qualifications do you have?"
	Income	Salary	"What is your monthly net income?"
		Interest income	"What is your annual interest income?"
	Occupation	Job/profession	"What is your profession?"
		Position at the job	"What position do you hold at your job?"

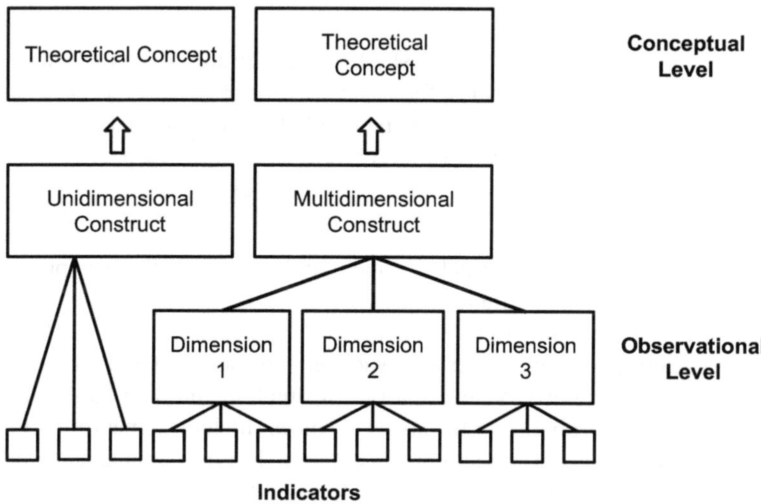

Fig. 6.1 Unidimensional vs. multidimensional operationalization of constructs

The **correspondence problem** relates to the assignment of suitable indicators to a theoretical concept (Wilson and Dumont 1968). It is about the "appropriateness" of an indicator for the concept to be measured (correspondence rules) and also about the question of which indicators from a large number of possible indicators should be selected (indicator selection). A solution to the correspondence problem is the equation of theoretical terms with prescribed measurement or observation instructions. For example, one can assume that intelligence is what an intelligence test measures. In the theory of science, so-called *operationalism* deals with such equations (Tal 2015; Trout 2000). However, one runs the risk of neglecting possible deviations from operationalization and concept: Intelligence is a very complex concept, and it is probably extremely difficult to find indicators that fully and accurately capture intelligence. An alternative is to consider deviations of an operationalization from the theoretical concept by modeling measurement errors, which should be as minimal as possible. This refers to the reliability of a measurement, which is a prerequisite for the validity of a measurement. Section 6.3 will explain this further, after the following section clears up the nature and function of measurements.

6.2 Nature and Function of Measurements

Measurement is the application of the selected methods to the corresponding parts of reality. This is a two-way process: subjects, objects, etc. are confronted with measurement instruments, and the measured values (data) flow back (see Sect. 5.2).

> Nunnally and Bernstein (1994, p. 3) present a definition for measurement: "Measurement consists of rules for assigning symbols to objects so as to (1) represent quantities of attributes numerically (scaling) or (2) define whether the objects fall in the same or different categories with respect to a given attribute (classification)."

For example, the Fishbein model (Fishbein and Ajzen 1975) gives precise rules on how to determine an attitude value and assign it to an entity. If a study investigates companies and wants to measure the size of a company, the scaling could be based on specific characteristics such as sales or number of employees, and then companies are assigned to the categories of small, medium, or large enterprises.

There are specific problems in the social sciences (including marketing research) with regard to the nature of measurements compared to measurements taken in the natural sciences (for example, in physics) (Chang and Cartwright 2008):

- Most of the measurements in marketing research are, to a certain extent, **indirect**; that is, we use verbal information (mostly collected through questionnaires) from managers, customers, etc. on the measures of interest (for example, product quality, sales growth, corporate culture). The problem of inaccurate or systematically distorted information in surveys is well known (see, e.g., Groves et al. 2009). Marketing research refers to these problems in regard to company surveys as "*key informant bias*." It primarily has to do with which kinds of people in companies should be surveyed in order to obtain information about the data of interest and also how meaningful the information of such key informants is. Typically, key informants provide less information about themselves than about the organization they belong to (such as sales, structures, work processes, or decision-making processes). The data collected in this way may be subject to errors, however. In a large-scale study, Homburg et al. (2012) systematically examined factors influencing the quality of information provided by key informants.
- Most measures are obtrusive and can lead to **reactivity**. Obtrusiveness of a measurement occurs when a respondent is aware of the act of measuring, which

is commonly the case in surveys and is especially true for laboratory experiments. If this results in an influence on the response behavior, reactivity occurs (Campbell and Stanley 1963). Such problems occur very often when measures refer to social norms (e.g., environmentally conscious behavior) or individual values (e.g., success).

- The applications of social science measurements are subject to specific limitations. On the one hand, the measurement tools must be relatively **straight-forward** because they should apply to a wide variety of respondents in a variety of situations. On the other hand, ethical standards apply, which include prohibiting the exposure of subjects to excessive stress or invading their privacy (see Sect. 10.2.3).

The aim of measurement is a structurally correct transfer of an empirical relationship into a numerical relationship so that the numerical relations reflect the actual existing relations. The empirical relationships between the measured values of a variable correspond to different mathematical relationships between numbers. We call this the **measurement,** or **scale**, levels, and they indicate which numerical information corresponds to the actual empirical information. There are four different levels:

- The *nominal scale level* contains information about an equivalence relationship; that is, whether identical or unequal characteristics of a variable exist (for example, gender with the expressions male or female);
- The *ordinal scale level* provides information about an ordered relationship ("more or less" or "smaller or larger") of the values of a variable (for example, social class with the characteristics lower, middle, and upper social class);
- The *interval scale level* allows interpretation of the distance between individual values of a variable (e.g., temperature in Celsius);
- The *ratio scale level* also allows for the interpretation of the ratio of two values of a variable by means of all arithmetically possible operations, as a clear zero point is defined (using income as an example, an income of 500 euros would be half of an income of 1000 euros and five times more than an income of 100 euros. An income of 0 is independent of the unit of measure, or currency; for a person who earns nothing, it does not matter whether she or he earns 0 euros, 0 pounds, or 0 dollars).

There is a hierarchical order between the scales or measurement levels, with the nominal scale representing the lowest measurement level and the ratio scale, the highest. All information of a lower measurement level also apply at a higher measurement level. However, the information of a higher measurement level cannot be used at a lower measurement level (see Fig. 6.2). In general, the higher the scale or measurement level, the more informative the measurement.

For the measurement of concepts or latent variables, as mentioned above, several indicators are typically used. The term **scaling** generally defines a scale for a variable. Mostly, it is about the construction of a scale that primarily measures a concept or a latent variable that consists of several indicators. Scaling refers, in

	Equivalence	Ranks	Interval	Ratio
Nominal scale	X			
Ordinal scale	X	X		
Interval scale	X	X	X	
Ratio scale	X	X	X	X

Fig. 6.2 Relationship of scales and their interpretation

particular, to the selection of suitable indicators or items and appropriate response options. For further use in analysis, the indicators used to measure a concept or a latent variable are usually combined into one index. An **index** is a new variable formed by mathematical operations of multiple indicators to represent the concept. Often, this is done by averaging or summing, such as when attitudes toward a product are measured with three indicators (bad/good, negative/positive, worthless/valuable), each measured on a scale of 1–7. The individual scores on these three indicators can be either summed up or averaged to create an index that describes the individual attitude score of a person. Index formation may also be weighted, such as in the context of measuring income as a dimension of social class (see example above), where labor income might be weighted more heavily than interest income. In such case, the labor income may be weighted accordingly (for example, twice as much as the interest income) before combining the two indicators.

The process of scale development is often very complex [see the corresponding scale development procedures in Churchill (1979) or Rossiter (2002)]. Therefore, once developed and established scales are often reused in research [see, e.g., the collection of established scales in Bruner's *Marketing Scales Handbook* (Bruner 2017)]. What is important for the use and establishment in science is that these scales and measurements meet certain quality criteria, which will be discussed in the following section.

In addition to the rather pragmatic argument of efficiency, other important aspects, such as those listed below, speak in favor of the development and repeated application of standardized measuring instruments (Nunnally and Bernstein 1994, pp. 6ff.):

- Greater *objectivity* of the measurements because they are not determined solely by individually developed measuring methods.
- Better possibilities for the realization of *replication* studies that are based on the same measurement methods (see Sect. 9.2).
- Easier and better *communication* of test results by reference to measurement methods that are known and recognized by experts and the academic community.
- *Comparability* (over time, across different groups or regions, etc.) of study results, which, given the strong influence of measurement methods on results in social science studies, can be more likely obtained by using uniform methods (Li 2011).

6.3 Quality Criteria for Measurements

6.3.1 Nature and Relevance of Validity and Reliability

The foregoing has already shown that translating theoretical concepts into measurable variables is not an easy task. However, study results are only meaningful if this implementation succeeds. The question of whether the translation of a theoretical problem into a study design (with sampling, measurement methods, etc.) is appropriate has the greatest importance for the value and merits of empirical studies.

As discussed in Sect. 5.2 above, there are two main problems. First, it is important to understand that a study, with all its methodological details, can lead to a systematic deviation from the "true value" of the examined concept. Second, random factors can influence study results during the measurement procedure. Sometimes, there is also the requirement that the measurement and results are independent of the influence of the investigators or the study situation. This is called **objectivity**. From a philosophy of science point of view, scientific knowledge can be dependent on contexts (see Sect. 3.2). Empirically, the objectivity of a measurement can be tested by, for example, generalizing the measurement across different researchers and different contexts (see Sect. 6.3.4). In the vast majority of the literature, however, objectivity is not treated as an independent aspect of measurements but rather as a partial problem of validity, because a lack of objectivity leads to a systematic distortion of the results.

The two aspects of systematic deviation of measurements and the influence of randomness lead to the two fundamental criteria for determining the quality of empirical research studies: Validity, which refers to the avoidance of systematic deviations of the test results from the true values, and reliability, which is about the independence of study results from one-time measuring processes and randomness.

- The **validity** of a test result can, thus, be characterized as follows: A study result is considered valid if it actually reflects the facts to be determined.
- **Reliability** can be characterized as a test result's high levels of independence from a one-time investigation and the ensuing random situational influences.

David de Vaus (2002) characterizes the relevance of reliability and validity as noted below:

Reliability: "If we cannot rely on the responses that a questionnaire item elicits then any analysis based on such data will be suspect. If the results we obtain from a sample could just as easily be different if we administered the questionnaire again, how much confidence can we have in any of the findings?" (p. 17)

Validity: "Since most social science analysis relies on using relatively concrete measures of more abstract concepts we face the problem of knowing whether our measures actually measure what we say they do. This is the problem of validity. We must somehow be confident that our relatively concrete questions actually tap the concepts we are interested in." (p. 25)

The relationship between validity and reliability can be illustrated by a simple formula based on Churchill (1979):

$$X_0 = X_T + E_S + E_R,$$

where $X_0 =$ measured, observed value; $X_T =$ the true (usually unknown) value of the concept to be measured; $E_S =$ systematic error in a measurement (e.g., by survey questions that favor a certain response tendency); and $E_R =$ random error in a measurement (for example, by situational factors such as time pressures that influence long-term and stable opinions, intentions, preferences, etc.)

A measurement is considered valid if there are no systematic and no random errors:

$$E_S = 0 \, \text{and} \, E_R = 0 \, \text{and therefore,} \, X_0 = X_T$$

From the reliability of a measurement, $E_R = 0$, it does not necessarily follow that the measurement is also valid, since $E_S \neq 0$ can be true. In this sense, *reliability is a necessary but not sufficient condition of validity.*

The fundamental importance of reliability and validity for empirical studies is obvious. If these requirements are not met, then the results of a study do not reflect the parts of reality that are of interest and, therefore, have no significance for the examined research question. The above statement that reliability is a necessary, but by no means sufficient, condition of validity is easy to understand when one considers that study results with low reliability are subject to strong fluctuations in repeated measures such that it represents a stroke of luck to meet the true value with sufficient accuracy.

The problems of lack of validity and reliability are illustrated by two examples that build on the basic model of empirical research described in Sect. 5.2. The model represents the relationship between reality, theory, and method. Theory and method are related on the one hand through operationalization and on the other hand through the interpretation of results.

In the first example, it is shown that by a (grossly) erroneous operationalization, measurements were made that do not correspond to the (theoretical) concepts of interest (X, Y instead of A, B); that is, they are not valid. The result is that the investigation says nothing about the question (A → B?). The problem is exacerbated by the fact that such measurement errors often remain undetected, and the result of the study is then misleadingly interpreted (despite its lack of informative value) with regard to the initial question.

(continued)

Example 1

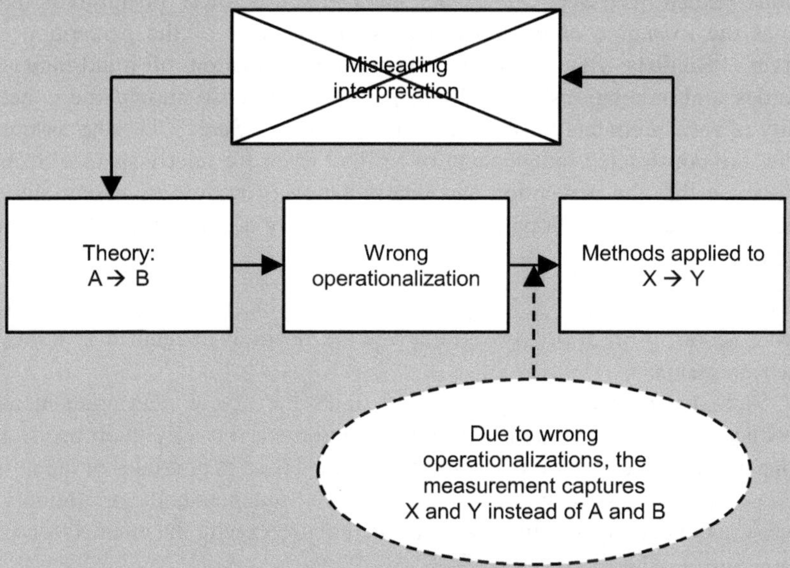

In the second example, it is shown that during the measuring process itself, a (non-systematic) error has occurred due to an outlier. This is an example of lack of reliability. The result is that the theoretically assumed linearly positive relationship between the two concepts is not reflected in the data or the study results, and the (actually correct) hypothesis is rejected, which is also a misleading result.

Example 2

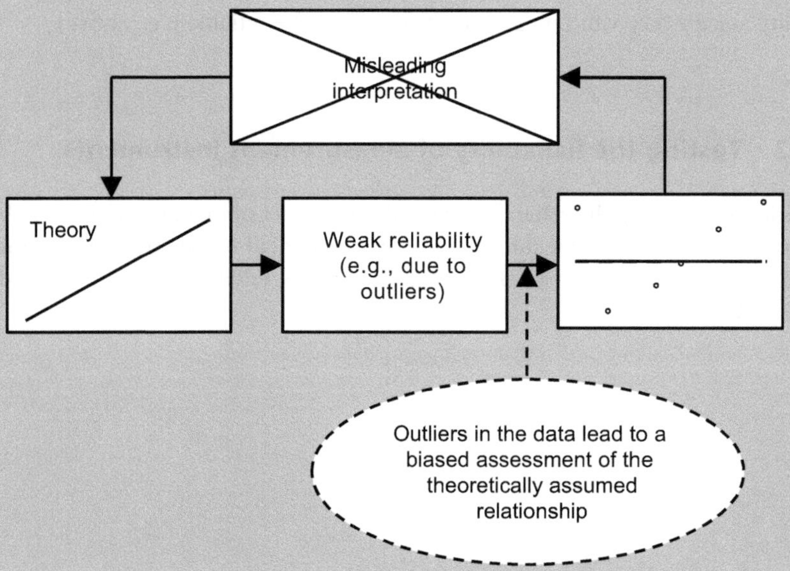

The reference of validity and reliability to a true value suggests a conceptual link to realistic philosophy of science positions (for example, critical rationalism or scientific realism; see Sect. 3.1). One characteristic of these positions is that one assumes the *existence of a reality* that is independent of the perception of the observer. "**Realists** view measurement as the estimation of mind-independent properties and/or relations" (Tal 2015, p. 2). What sense should the concept of validity of measurements have for constructivists (see Sect. 3.1), who assume that theories are constructed independent of reality? Even for relativists (see Sect. 3.1), who assume that the perception and interpretation of reality are essentially determined by a social context or paradigms, the possibility of achieving validity is hardly a given.

Trout (2000, p. 272) briefly summarizes the position of realism in terms of measurements:

"A realist account of measurement treats the act of measurement as a product of a causal relation between an instrument (broadly interpreted) and a magnitude. The relation is one of estimation. These magnitudes or quantities (properties, processes, states, events etc.) exist independently of attempts to measure them, and are sometimes too small to detect with the unaided senses."

To conclude these considerations, the relevance of validity and reliability from the perspective of scientific realism is explained by referring to the *inductive-realistic model of theory testing* presented by Hunt (2012). In Fig. 6.3, it is easy to see the relations that each stand for the correspondences of entities, attributes, and relationships in theory and reality (external world). If these correspondences are given, that is, if the entities, attributes, and relationships in theory coincide with the measurements of entities, attributes, and relationships in reality (external world), then one speaks of validity. Reliability is symbolically included as a requirement of validity since every valid measurement must also be reliable (see above).

6.3.2 Testing the Reliability of Measurement Instruments

Criteria for assessing the reliability and validity of measurements will be only briefly described here, since the technical and methodological details of such methods are not the subject of this book. This section examines reliability, and then validity is discussed in Sect. 6.3.3.

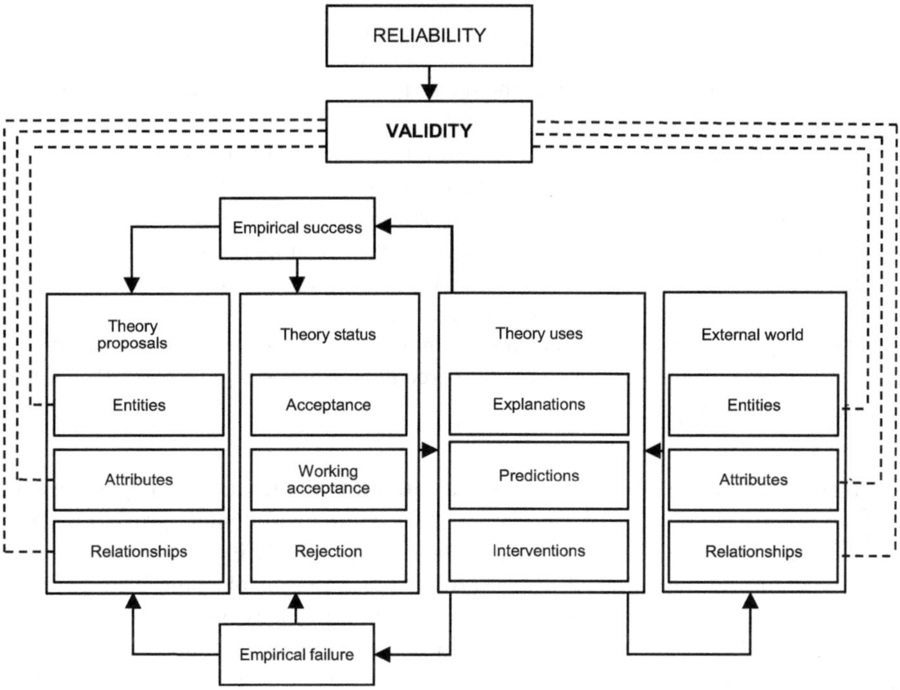

Fig. 6.3 Validity and reliability in the (simplified) inductive-realistic model of theory testing (Hunt 2012)

Peter (1979, p. 166) explains the basic ideas of reliability theory:

"[The] basic approach starts with the notion that the mean and variance of any observed scale score can each be divided into two parts. In terms of the mean, the two parts are the true score and the error score or

$$X_{\text{observed}} = X_{\text{true}} + X_{\text{error}}$$

Conceptually, the true score is a perfect measure of the property being measured. However, in practice, the true score can never really be known and generally is assumed to be the mean score of a large number of administrations of the same scale to the same subject. The error score is an increase or decrease from the true score resulting from measurement error. Measurement error is the source of unreliability and its primary cause is that items in the scale are not measuring the same phenomenon.

The variance of an observed scale score also is assumed to have a true component and an error component or

(continued)

$$V_{observed} = V_{true} + V_{error}$$

The true variance component includes all systematic variance. In one sense, it is a misnomer because it includes both from the phenomenon under investigation and all other sources of systematic variance. (Determination of the difference between types of systematic variance is a validity question.) The error variance component includes all random or nonsystematic variance. In terms of the previous definition of reliability, systematic variance does not affect either the rank order or distance between subjects but random or error variance does and thus error variance lowers the reliability of measures. A reliability coefficient ... therefore, is nothing more than the ratio of true variance to observed variance or the percentage of total variance which is of the systematic type."

When assessing reliability, the fact that **reliability** relates to the independence of the measured values from the randomness of a single measurement process is taken into account. The basic idea of so-called **test–retest reliability** is directly linked to it. It is about the repetition of a measurement at a reasonable time interval. As a measure of reliability, we would use the correlation of the two measurements. This type of reliability check assumes that the observed construct has not changed in the meantime; otherwise, a low correlation would not be due to a lack of reliability but, rather, to this change. Reliability testing by repeating a measurement and comparing the results is quite laborious. This can be avoided with the use of **parallel-test reliability** by carrying out a comparison measurement with a different, but equivalent, measurement instrument at the same time (usually in the same questionnaire). Both measurements should be highly correlated in order to provide evidence for their reliability. The difficulty here is to find or develop two *equivalent* measurement instruments.

Probably the most common type of reliability check is the determination of the reliability coefficient **Cronbach's α** for a multi-item scale, meaning for the measurement of a latent variable that uses several items or indicators (Cronbach 1951). It is a measure of the internal consistency of a scale, or the degree of agreement of the measurement values for each indicator of a scale. It is assumed that all items or indicators measure the same true value and only random measurement errors lead to different results. Cronbach's α is calculated as a corrected average correlation between the items or indicators. Cronbach's α can maximally take on the value 1; a higher positive value proves higher internal consistency. Values below 0.7 fall into the category of "questionable". It should also be noted here that Cronbach's α can be influenced (improved) in various ways, which may well raise ethical research questions (see Sect. 10.2.2). For example, Cronbach's α increases with the number of indicators or items used (Peterson 1994).

6.3.3 Testing the Validity of Measurement Instruments

At the center of interest in the development and verification of measurement instruments is their **validity**. With the validity stands and falls the quality of a measurement and, thus, of the whole study in which it is used. The central term—commonly used in literature—is **construct validity**. We use "construct" and "concept" as synonyms. Construct validity thus signifies the correspondence of a theoretical (and usually not directly observable) concept/construct with a corresponding measurement.

Nunnally and Bernstein (1994, p. 84) characterize the relevance of construct validity:

"All basic sciences ... are concerned with establishing functional relations among important variables. Of course, variables must be measured before their interrelations can be studied. For such statements of relationship to have any meaning, each measure must validly measure what it purports to measure."

Typically, we cannot determine the validity of a measurement by comparing the measured value with the true value of the concept of interest because that value is usually unknown, although it may be possible in a few exceptional cases. As a rule, however, a tedious method of operationalization and measurement is necessary because the desired data of true values are not available. Often, empirical research is about concepts such as satisfaction, attitudes, or intentions, where a true value cannot be determined. In these cases, different auxiliary criteria are used to determine whether the measurement method corresponds to different types of validity. These are

- Content validity
- Criterion validity
- Convergent validity
- Discriminant validity

If a measurement method survives the different types of validity testing, it strengthens confidence (in the sense of scientific realism) that this method actually measures what it should measure. From there, we can draw scientific conclusions based on the resulting findings, though we can never be absolutely sure about them. Importantly, we always assume the reliability of the corresponding measurements (see above). The reader may keep in mind the relevance of measurements for the validation/falsification of theories and explanations, as detailed in Sect. 5.2.

First, the **content validity** refers to the suitability and completeness of the measurement instrument (usually judged by experts) with regard to the observed concept or construct. The point here is that the essential aspects of this concept are reflected in the question formulation or in the various items of a scale. The essential content must be derived from the definition of the concept, and the measurement instrument must include it.

David de Vaus (2002, p. 28) gives an example of content validity:

"Assessing content validity involves examining the extent to which the measure taps the different aspects of the concept. For example, a measure designed to gauge general health that confined itself to blood pressure would not adequately tap the concept of health—not, at least, as it would normally be understood. Health would usually be understood to be something much broader and more complex. Other aspects of physical health as well as, for example, psychological health would normally form part of a valid measure."

The possibilities for checking the **criterion validity** are much more concrete. Criterion validity refers to the fact that the result of a measurement has a known (established) relationship with measurements of other concepts. For example, it has long been known in behavioral research that attitudes and behaviors have a (non-deterministic) positive relationship. If one develops a scale for measuring environmental protection attitudes (high value = positive), then the resulting values would have to be positively correlated with environmental behavior measurements (e.g., waste separation). Otherwise, the validity of the attitude scale would be doubtful. Figure 6.4 illustrates the basic ideas of testing content and criterion validity.

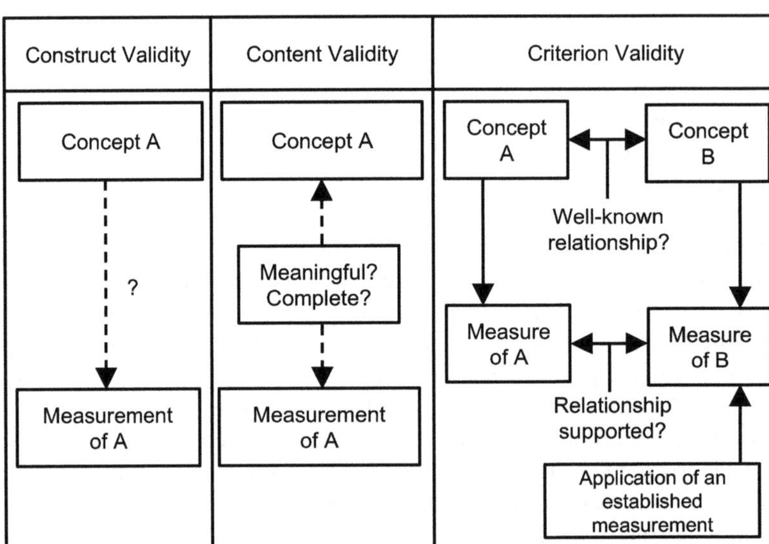

Fig. 6.4 Basic ideas of testing content and criterion validity (source: Kuß and Eisend 2010, p. 102)

Central to validity testing are convergent validity and discriminant validity. First, with regard to **convergent validity**, if the same concept is measured by two different measurement instruments that are both valid, then the results should be similar (converging). *Both* instruments should have as few methodological similarities as possible; otherwise, the similarity of the measured values could be an artifact caused by just these similarities. Thus, if two *dissimilar* measurement methods applied to the same concept lead to convergent results, then these results appear to be independent of the measurement approach, which in turn supports (but, of course, does not prove) that the measurement methods reflect the concept of interest.

What is the central idea of **discriminant validity**? If one measures *different* (unrelated) concepts with the same type of measurement instrument (e.g., Likert scales), then the results should not be correlated. Otherwise, the measurements would not reflect the difference in concepts, but rather, would be due to systematic influences of the measurement methods, which of course, would lower our confidence in their validity. In the case of similar measurement methods applied to different concepts, the measured values for these concepts should clearly differ (discriminate). Figure 6.5 illustrates the basic ideas of both approaches.

Convergent and discriminant validity can be assessed empirically by using the **multitrait–multimethod matrix** (Campbell and Fiske 1959). Figure 6.6 shows a schematic representation of a matrix with two concepts (A and B), each of which is measured by means of two measurement instruments (M1 and M2; for example, two different scales). In the matrix itself, the correlations **r** between these measurements are shown. The letters C and D behind the correlation coefficients indicate which correlation coefficients are decisive in terms of convergent and discriminant validity. The arrows next to them indicate whether the correlations should be high or low to confirm convergent or discriminant validity. In particular,

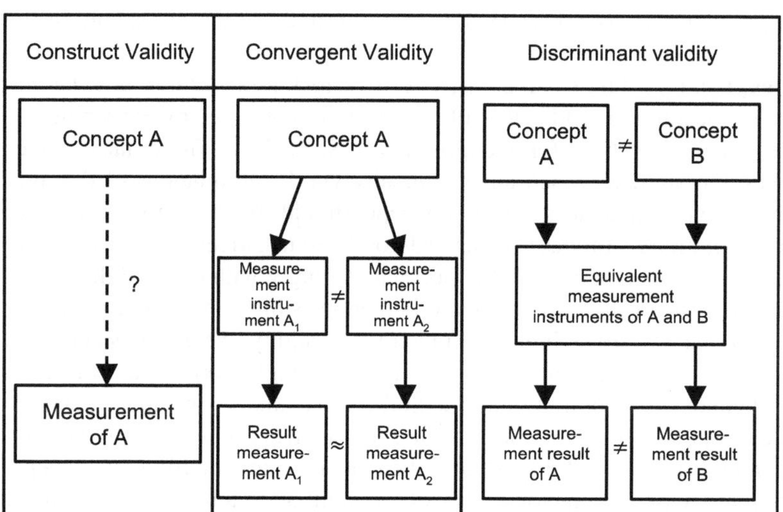

Fig. 6.5 Testing convergent and discriminant validity (source: Kuß and Eisend 2010, p. 103)

		M_1		M_2	
		C_A	C_B	C_A	C_B
M_1	C_A				
	C_B	$r_{AB,11}$ (D↓)			
M_2	C_A	$r_{AA,12}$ (C↑)	$r_{AB,21}$		
	C_B	$r_{AB,12}$	$r_{BB,21}$ (C↑)	$r_{AB,22}$ (D↓)	

Fig. 6.6 Multitrait–multimethod matrix (source: Kuß and Eisend 2010, p. 104)

- The coefficients $r_{AA,12}$ and $r_{BB,21}$ show how strongly the values measured by *different methods* for the *same concept* correlate. If the correlations are high, convergent validity exists. The correlations are expected to be significantly higher than the correlation coefficients used to test discriminant validity.
- The coefficients $r_{AB,11}$ and $r_{AB,22}$ show the correlations of measured values for *different concepts* that were measured by *identical measurement instruments*. If, as is assumed, there is no relationship between the concepts, and the corresponding measurement instruments correctly measure the concepts, then the correlation coefficients would have to be very small.

Nowadays, reliability and validity criteria, which are based on exploratory and confirmatory factor analyses, are also used. They are briefly outlined here; further details, particularly technical details, can be found in the corresponding literature (e.g., Netemeyer et al. 2003).

In **exploratory factor analysis**, one tries to find structures in a larger set of variables, meaning the indicators of one or more concepts, by extracting so-called factors (Jackson 1969) (for examples, see Sect. 7.6). These factors are latent variables and are determined algorithmically via the correlations of the indicators. Ideally, several indicators for a concept should be strongly correlated (e.g., indicators of brand attitudes such as bad/good and negative/positive should strongly correlate). They will then probably correlate strongly with one and the same factor, and on the basis of these indicators, it is then possible to extract exactly one factor (here, attitude toward a brand). Convergent validity therefore occurs when the indicators used to measure a concept are all strongly correlated with a common factor. Unlike the multitrait–multimethod matrix, this is not about measuring the same concept with different scales but about *convergence of the indicators for a concept.* Discriminant validity occurs when the indicators of different concepts result in different factors in a factor analysis (e.g., attitude to a brand versus life satisfaction). Ideally, the indicators correlate with only one factor at a time, meaning the respective indicators that relate to a particular concept can be clearly assigned to one of the factors.

In **confirmatory factor analysis**, in contrast to exploratory factor analysis, there is already a theoretical assumption as to which indicators can be assigned to which concepts. The confirmatory factor analysis is also referred to as a measurement model that is to be verified on the basis of empirical data (Bagozzi 1978). Again, convergent validity occurs when the indicators are strongly linked to the respective factor. In contrast, the correlation between different factors related to different concepts should be as low as possible to confirm discriminant validity. Fornell and Larcker (1981) have proposed a criterion based on two measures:

- The *average variance extracted* of a construct is a measure of how well a single latent variable explains its indicators. In confirmatory factor analysis measurement models, one usually assumes that the indicators are explained by the concepts/constructs/latent variable (e.g., the statement "I am generally satisfied" is explained by the construct "life satisfaction"). However, the explanation is not perfect, and an error term remains (see also Sect. 7.5). An indicator is, thus, explained by the latent variable and the error variance.
- The *squared correlation* between the constructs means the correlation between the constructs (e.g., between "attitude toward a brand" and "life satisfaction") and measures the strength of the relationship between these constructs.

If the average variance extracted of a construct is higher than any squared correlation with another construct, it is evidence for discriminant validity. That is, one construct explains more variance of the associated indicators than variance of another different, unrelated construct. This measure of quality is called the **Fornell–Larcker criterion**. The average variance extracted should also be large enough to provide evidence of convergent validity. Ideally, more than 50% of the variance of each indicator should be explained by the construct, or at least half of the overall variance of all indicators are explained by the construct (and, thus, are greater than the error variances) (Hair et al. 2010).

Another validity criterion is the **nomological validity**. This refers to the confirmation of theoretically suspected (causal) relationships of one variable to several other variables. Here, a logical problem arises when a measurement is to be used to test a theory and the confirmation of the relationships with other variables within this theory are used as criteria of validity (see Nunnally and Bernstein 1994, pp. 91–92) The nomological validity could, thus, be tested with respect to a nomological network that is not identical to the theory that ultimately should be tested. For measurements that do not aim to test a theory (instead testing, for example, a practical application), the criterion of nomological validity is directly applicable.

It has been shown that there is no direct way to confirm (or even prove) construct validity. Instead, various tests are used (→ content validity, criteria validity, convergent validity, discriminant validity) that do not provide any *proof* of validity but allow a critical review of validity. This is in the spirit of scientific realism, and

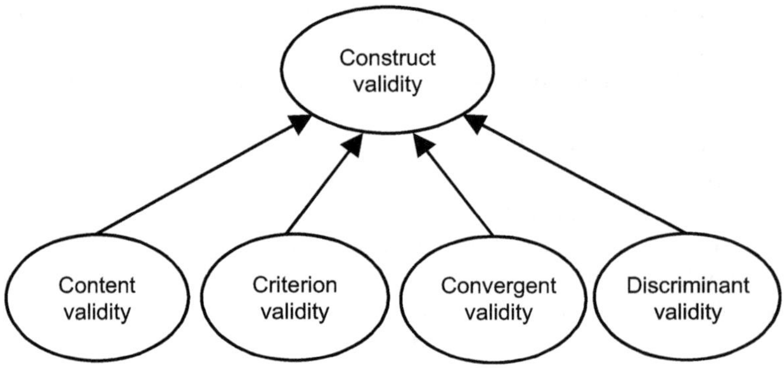

Fig. 6.7 Criteria to assess validity (following Viswanathan 2005, p. 65)

positive results give reason to believe that a measuring instrument meets the requirement of construct validity. Figure 6.7 illustrates this idea.

6.3.4 Generalizability of Measurements

In addition to reliability and validity, there is the requirement of *generalizability* or *transferability* of measurements. The aim is that measuring instruments should be applicable in different contexts, to different persons, or at different times, and the results should be comparable across contexts, persons, and times. Two ways to assess and guarantee generalizability of measures are presented next. The concept of dependability of measures, or generalizability theory, aims at the generalizability of measurements with respect to different dimensions. The idea of measurement invariance or equivalence also refers to the applicability and transferability of measurements to different groups of respondents or to different contexts but is typically used in marketing research in the application of measuring instruments in different cultural contexts.

The concept of *dependability* of measures is directly related to reliability. The above-mentioned reliability tests focus on the influence of different random sources of error. Thus, test–retest reliability measures the influence of the error source time, parallel-test reliability measures the errors that are caused by subjects or participants, and the reliability coefficient α measures the errors attributable to different items or indicators. The lower the measurement error, the better the measurement can be generalized over one of the corresponding conditions (time, subject, items) insofar as the reliability tests also make a statement about the generalizability of a measurement. However, measurement errors can be based on *different sources of error at the same time* whereby these different sources of error can also influence one another. Conventional reliability tests cannot take this problem into account. A consideration of different sources of measurement errors that can occur simultaneously, as well as their interactions, is provided by **generalizability theory** (Cronbach et al. 1972). Figure 6.8 illustrates the relationship between reliability tests and generalizability. It

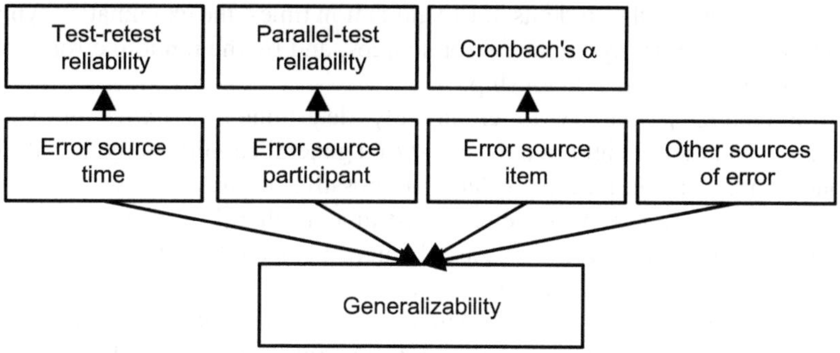

Fig. 6.8 Relationship between reliability tests and generalizability theory

should be emphasized that these considerations have nothing to do with the validity of measurements. The only question here is the extent to which results can be transferred to other points in time, participants, etc., irrespective of whether the results are valid with regard to the respective constructs.

The generalizability theory is based on a statistical procedure that is significantly more complex than other methods of reliability testing and is here only explained in its basic features. Detailed descriptions can be found, for instance, in Cronbach et al. (1972), Brennan (2001), Rentz (1987) or Shavelson and Webb (1991). As an example, the concept of student performance, which is measured by exams, will be used. Student performance is a complex and abstract concept where there is also a legitimate (and easily comprehensible) interest in the results of the measurement being as free of measurement errors as possible.

Fundamental to this is the assumption that each observed value of a study object x_i (e.g., the exam performance of a student) represents a sample of a universe of possible observations under different conditions y, z, ... (e.g., examination time [morning, noon, evening], type of examination [oral, written]). The expected value of a study object across all these observations is called the **universe score** or **global true value**, which is the true value in the sense of the reliability tests (in the example above, it would be the student's true performance on an exam). The universe to which the measurement is to be generalized is determined by the researcher on the basis of theoretical considerations, using the characteristics of generalization that seem to him or her to be important. In the above example, the performance of students is to be generalized over different examination times and different types of examinations; that is, these two dimensions are taken into account as possible sources of error.

Analogous to the idea of reliability tests, an observed value is then composed of the so-called universe score and an error term. Other than in reliability tests, the error term can be broken down into several components. For this purpose, we use the method of analysis of variance, with which an observed value can be decomposed into different variance components, which are due to the effects of independent variables, their interactions, and an error term. Hence, a measured value in the given example can have various influences (so-called *variance sources*): the measured value is influenced by the students, the examination times, the examination types and their interactions, other systematic error sources, and by the random error variance. Figure 6.9 illustrates this relationship.

In the next step, we try to empirically determine the *individual variance components* and their weight in the context of a generalizability study. It is desirable that one source of variance is as large as possible because we want to explain variance by applying a measurement instrument; that is, we want to explain differences between the study objects of interest. In the given example, students would be the source of variance that we want to be large because we want to measure performance differences between the students. All other sources of variance should be as minimal as possible because we want to avoid having the performance of the students depend on coincidences such as the time of examination or the type of examination. One can, therefore, make a distinction between **facets of differentiation** and **facets of generalization**. The facet of differentiation refers to the actual object of investigation, in our example, the students, while the facets of generalization represent the sources of error of the measurement.

One aim of a generalizable measurement is to make the variance of the facet of differentiation as large as possible in relation to the variance of the facets of generalization: the students' performances may vary, but they should not depend on the time

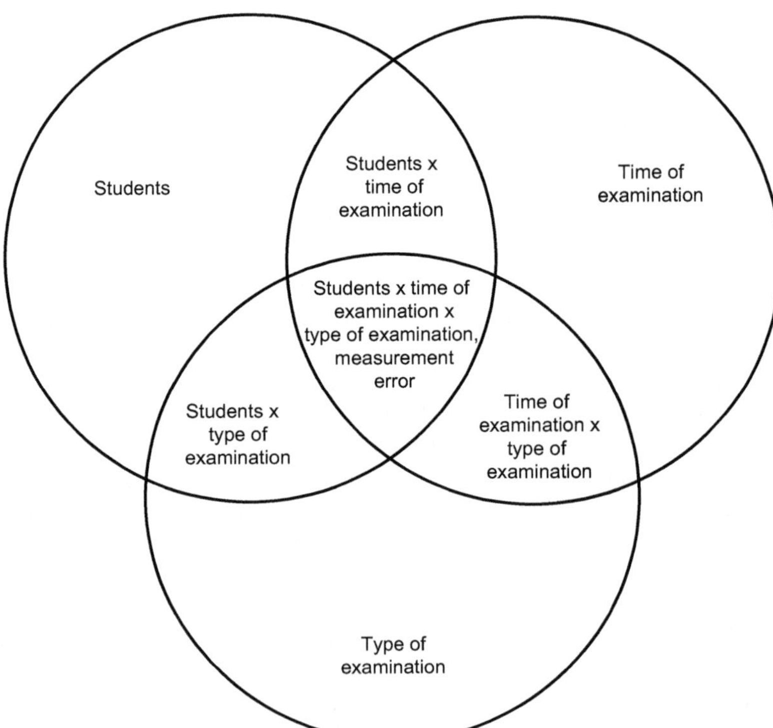

Fig. 6.9 Sources of variance (students, time of examination, type of examination) and their interactions in measuring performance

of examination and the type of examination. The determination of the facets is dependent on the study's purpose and can also vary. For example, when measuring customer loyalty, customers can be the facet of differentiation, while different brands are the facets of generalization. However, the individual brands can also be seen as a facet of differentiation in order to be able to differentiate between different brand concepts and the associated brand loyalty of consumers (Rentz 1987).

In addition to an **exploratory generalizability study** that identifies and quantifies the variance components, the next step requires a decision study that works with the components making an important contribution to explanation of the variance, thus attempting to optimize the design of an appropriate measurement (in this example, exam performance); that is, to increase its generalizability. Facets of generalization, which only make a small contribution to the total variance, can also be completely ruled out because we can, obviously, assume that there is sufficient generalization. For example, if the variance component of the type of examination is low, we can assume that different types of exams have no impact on exam performance.

The **decision criterion** is analogous to the reliability coefficients and refers to so-called *generalizability coefficients*. If, for example, the time of the examination has a high level of variance, this variance can be taken into account through the extension of the different examination times, and the generalizability of the measurement can be increased. The result is a study design that shows how to design the facets of generalization to achieve greater generalization. In the example, it might be that students are to be examined at all possible examination times (morning, noon, evening) because their performances might depend on the examination times. By considering all potential examination times, we can minimize this source of error.

When applying the generalizability theory, for practical reasons, only a limited number of sources of error can typically be investigated, and therefore, its application requires a random selection of components. This not inconsiderable effort and methodological complexity probably contribute to the fact that applications of Cronbach's generalizability theory, in contrast to Cronbach's α, can seldom be found in the practice of test construction.

The second approach to assessing the generalizability of measures is the concept of **measurement invariance**, or **measurement equivalence**. This is commonly used in *cross-cultural research* when abstract concepts measured with several items are applied in different cultural contexts (e.g., Steenkamp and Baumgartner 1998). Measurement invariance refers to "whether or not, under different conditions of observing and studying phenomena, measurement operations yield measures of the same attribute" (Horn and McArdle 1992, p. 117). Typically, a construct measurement is developed and tested in a particular cultural context. If the measurement is applied in different cultural contexts, the question arises whether differences in findings do, indeed, represent cultural differences or if they are subject to systematic biases in the way people from different countries respond to particular questions. The idea is similar to the one of generalizability theory in that it tries to disentangle various sources of variance (e.g., variation due to cultural differences or due to measurement errors). Obviously, the question applies not only to cultural differences, but can also apply to groups of people of different ages or gender, or differences in other contexts such as study design, time of measurement, etc. To assure measurement invariance, the relationship between each item and the concept it represents should be equivalent across groups.

This essentially refers to how multi-item scales are modeled in *confirmatory factor analysis* (Steenkamp and Baumgartner 1998). The response to an item, x_i, is represented as a linear function of a latent variable, ξ_j. The equation further includes an intercept, τ_i, and an error term, δ_i:

$$x_i = \tau_i + \lambda_{ij}\xi_j + \delta_i,$$

where λ_{ij} is the slope of the regression of x_i on ξ_j. The slope coefficient corresponds to the factor loading and shows the degree of change in x_i due to a unit change in ξ_j (i.e., the metric of measurement). The intercept τ_i refers to the expected value of x_i when $\xi_j = 0$. Of course, a multi-item scale consists of several items and, therefore, a set of equations applies.

Measurement invariance refers to different forms of invariance that relate to different elements of the above formula. The forms of invariance are sorted along the strength of the test, meaning that the preceding requirement is weaker than the following ones. *Configural invariance* refers to the loading $\lambda_{ij}$of each item. To ensure measurement invariance, the loading should be substantially different from zero. If this requirement is not met, the items are assigned in different ways to a construct or its dimensions, and as a result, different concepts are compared across countries, and the measurement results cannot be used to assess differences between countries. *Metric invariance* goes a step further and requires that the loading λ_{ij} is the same across groups (countries). If the loading is different, the structural relationships between items and the construct are biased across countries. *Scalar invariance* requires that the intercept τ_i is the same across countries. If this requirement is not met and the intercepts are different (while the loadings are equivalent), the interpretation of differences in means is biased. *Error variance invariance* requires that the amount of measurement error is invariant across countries. If this is not the case, reliabilities of measures do differ across countries, and measurement artifacts can hinder the comparison across countries. *Factor covariance invariance* is a requirement for the factor covariances (i.e., the covariation of several dimensions of a construct). If a construct has several factors, their covariances should be the same across countries.

The concept "attitude toward the brand" (ξ) is usually measured with multiple items, such as bad/good (x_1), unfavorable/favorable (x_2), and negative/positive (x_3). The corresponding equations for each of the three items that depend on the concept are as follows:

(continued)

$$x_1 = \tau_1 + \lambda_1 \xi + \delta_1$$

$$x_2 = \tau_2 + \lambda_2 \xi + \delta_2$$

$$x_3 = \tau_3 + \lambda_3 \xi + \delta_3$$

If we want to assess measurement invariance or equivalence across different countries, we proceed as follows:

Configural invariance: We assess whether λ_i is significantly different from zero in each country. That is, whether each item can be explained by the concept "attitude toward the brand" in each country. If this is not the case, the number of items needing to be measured in different countries would vary, and hence, the concept would not be generalizable.

Metric invariance: We test whether λ_i is the same across countries; that is, whether the concept explains an item to the same extent in each country. If not, the item's relationship to the concept varies across countries, and the measure would not be generalizable.

Scalar invariance: We test whether τ_i is the same in each country. That is, we test whether the means of an item are the same in each country. If not, consumers in a country systematically rate an item higher or lower than in other countries.

Error variance invariance: We test whether δ_i is the same in each country. That is, we test whether the measurement error is the same for a particular item in each country. If not, the measurement error is not invariant across countries.

Factor covariance invariance does not apply to this example since we have only one factor. If a concept has two factors (e.g., credibility is a two-factor concept that includes the dimensions of competence and trustworthiness), we will expect that the correlation between both factors will be the same across different countries.

As the above description shows, the different types of measurement invariance in marketing research are typically tested in the context of confirmatory factor analysis. The analytical procedure is described in more detail in the respective literature (Steenkamp and Baumgartner 1998; Vandenberg and Lance 2000). Essentially, the measurement model is compared across groups (e.g., countries) and the respective component (e.g., the intercept, or the slope coefficient) is constrained to be equal across groups. Ideally, the restriction should not affect the fit of the measurement model, and the fit should be comparable to a model without restrictions. Because full measurement invariance is not very common in practice, researchers usually try to achieve at least *partial measurement invariance*. This can be meaningful, depending on the goal of the study (Steenkamp and Baumgartner 1998). For instance, configural invariance is necessary if one wants to explore the basic meaning and structure of a construct cross-nationally. If one wants to compare means across countries, metric and scalar invariance is required. When researchers relate the construct to other constructs in a nomological network and try to compare the strength of association between constructs (e.g., the effect sizes) across countries, factor covariance invariance is necessary in addition to metric invariance. The scale reliabilities should be about the same, as well, thus requiring error variance invariance.

6.4 Data Collection and Sampling

Once valid, reliable, and generalizable measuring instruments have been developed, researchers can apply them to collect data from the units (subjects) they are interested in. Since researchers are interested in making broad and generalizable statements, the units they are interested in are often numerous as well, and it is not uncommon for researchers to try making general statements that apply to all consumers, companies, or human beings on the planet. As a result, it is frequently not feasible, or even possible, to collect data from the total of all units of interest (e.g., the population) and therefore, researchers collect data from a **sample**. "A sample is a proportion or subset of a larger group called a population ... A good sample is a miniature version of the population of which it is a part—just like it, only smaller" (Fink 2003, p. 1). Ideally, the results that a researcher obtains by collecting data within a sample should represent the results that the population would provide. In other words, the sample should be representative of the population, and the sample findings should be generalizable to the population the sample represents.

Sampling is the process of selecting a subset of units from within the population. There are several techniques for doing this that are broadly categorized as probability and non-probability sampling. *Probability sampling* relies on a random, or chance, selection method so that the probability of selection of population elements is known. In *nonprobability sampling*, the probability of selection of population elements is unknown. **Representativeness**, and hence, generalizability, requires that all units have either the same or an a priori known chance for being selected. This can—in theory—only be assured through random sampling, which is by means of probability sampling. While non-probability sampling cannot depend upon the rationale of probability theory, and respective statistics cannot be computed, the application of *quota sampling* (a nonprobability sampling technique) is quite common in market research (Battaglia 2008). Results based on quota sampling can lead to predictions (e.g., in election polls) that are as good as the ones derived from random sampling studies. This might seem surprising, but in practice, random sampling suffers from several problems, such as **non-response bias** (i.e., a particular group of people in the selected sample does not respond, and these people differ from those who do respond).

Empirical results in a sample (e.g., the average age of the sample participants) vary from sample to sample and are typically close to the true value in the population, but not necessarily identical. The difference between the sample value and the true value in the population is called the **sampling error**. If researchers apply a random sampling technique, they are able to estimate the sampling error, meaning how closely the sample represents the larger population from which it was drawn.

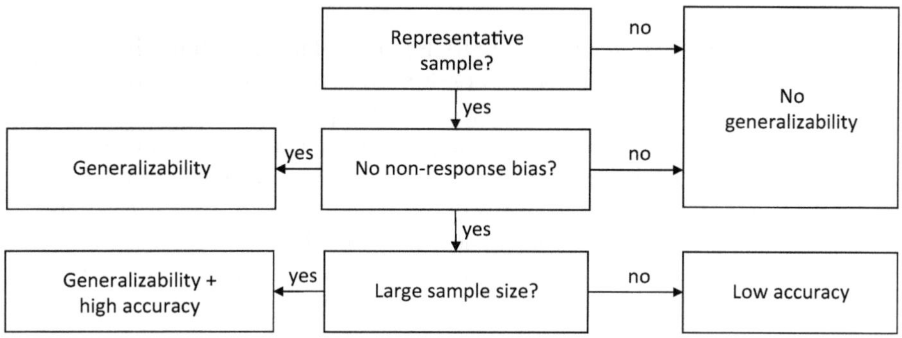

Fig. 6.10 Representativeness, generalizability, and accuracy

The sampling error plays an important role in hypothesis testing (see Chap. 7) because the general goal of a hypothesis test is to rule out chance (sampling error) as a plausible explanation for the results of a study. For instance, if a researcher hypothesizes that two different advertisements affect consumers in different ways, and the results from a sample of consumers reveals such a difference, the difference can be explained by the sampling error, by an actual difference between the two advertisements, or by both sampling error and differences in the advertisements. Therefore, researchers try to minimize the sampling error to increase the **accuracy** of test results. The sampling error *depends on the sample size and the variability* of some aspect within the population. While researchers cannot influence the variability in a population, they can alter the sample size. The literature provides procedures for computing the optimal sample size given a particular test power and effect size (Cohen 1992; see Sect. 7.3). Figure 6.10 illustrates the relationship between representativeness of a sample, its generalizability, and the accuracy of test results.

Sample size has become an important topic in research given that the internet has provided researchers with the opportunity to conduct studies using extremely large samples of well over 10,000 observations. Such **big data** obviously reduces the problem of sampling error. At the same time, big data inflates the number of significant findings because, in very large samples, even very small effects become statistically significant. With too much data, every difference is statistically significant because of too much test power increasing type I errors (i.e., a nonexisting effect is found to be statistically significant; see Sect. 7.3). Researchers have, therefore, started to question the meaningfulness of statistical tests in the era of big data and emphasize that *substantial significance* or effect sizes should replace significance tests (Szucs and Ioannidis 2017). Another caution regarding big data is that representativeness and generalizability cannot be traded for sample size. Even if a sample is very large, if it is not representative, any inferences about the whole population are not generalizable and are, therefore, misleading (Parks 2014).

The problem that researchers face in trying to reject a null hypothesis (i.e., the statement that no difference or relationship exists between two concepts) in a very large sample is illustrated by Cohen (1990, p. 144) as follows:

"A little thought reveals a fact widely understood among statisticians: The null hypothesis, taken literally (and that's the only way you can take it in formal hypothesis testing), is always false in the real world ... If it is false, even to a tiny degree, it must be the case that a large enough sample will produce a significant result and lead to its rejection. So if the null hypothesis is always false, what's the big deal about rejecting it?"

Experimental research in marketing often works with *convenience* (i.e., non-probability) *samples* of college students. In terms of representativeness and generalizability, such samples are obviously questionable because students are, in many ways, different from the whole population of consumers, decision-makers, managers, etc. These differences may bias their perceptions, evaluations, preferences, and behaviors, and hence lead to biased experimental results. However, as described above, researchers try to generalize their findings over time and to broader populations that are often not well described. Therefore, it is very difficult, if not impossible, to draw a random sample. Furthermore, student samples increase the internal validity of experiments (see Sect. 8.3.2) because of the homogeneity of the study participants. Students are homogenous on several dimensions (e.g., age, education, certain values), and this *homogeneity decreases variability in measurements* and increases the likelihood of rejecting a null hypothesis of null difference; that is, it helps in identifying theory violations if a theory is false. This is why researchers argue that college students constitute an appropriate sample when the research emphasis is theoretical and focuses on basic psychological processes or human behaviors independent of sample characteristics (Kardes 1996). Calder et al. (1981) distinguish between "*effects application*" and "*theory application*," with the former aiming at statistical generalization of a theory and the later aiming at theory confirmation. Statistical generalization requires that the research sample is representative of the population, and a student sample would be inappropriate unless it is representative for the population (e.g., in a study about binge drinking of students). However, in most cases representativeness, generalizability, and external validity (see Sect. 8.3.2) are at stake, because a student sample usually does not sufficiently represent some larger population of consumers or managers (Peterson and Merunka 2014).

Marketing researchers increasingly make use of samples drawn from crowdsourcing websites, in particular, *Amazon's MTurk* (www.mturk.com). The problem of lack of representativeness applies in a similar way here, because the MTurk respondents tend to be younger and more educated and tend to have a lower income than the general population. Several of these respondents have extensive experience participating in research studies or are more interested in maximizing their pay rate instead of answering thoroughly. Researchers therefore recommend taking precautions to ensure the quality of the data (e.g., by inserting attention check questions or checking the time respondents take to answer questions). With these quality checks in place, crowdsourcing samples can perform equally well or even better than other online or offline samples (Kees et al. 2017).

References

Bagozzi, R. P. (1978). The construct validity of affective, behavioral, and cognitive components of attitude by analysis of covariance structures. *Multivariate Behavioral Research, 13*(1), 9–31.

Battaglia, M. P. (2008). Nonprobability sampling. In P. J. Lavrakas (Ed.), *Encyclopedia of survey research methods* (pp. 524–527). Thousand Oaks, CA: Sage.

Brennan, R. L. (2001). *Generalizability theory*. New York: Springer.

Bridgman, P. W. (1927). *The logic of modern physics*. New York: MacMillan.

Bruner, G. C. (2017). *Marketing scales handbook. Multi-item measures for consumer insight research* (Vol. IV). North Charleston, SC.

Calder, B. J., Phillips, L. W., & Tybout, A. M. (1981). Designing research for application. *Journal of Consumer Research, 8*(2), 197–207.

Campbell, D. T., & Fiske, D. W. (1959). Convergent and discriminant validation by the multitrait-multimethod matrix. *Psychological Bulletin, 56*(2), 85–105.

Campbell, D. T., & Stanley, T. D. (1963). *Experimental and quasi-experimental designs for research*. Chicago: Rand McNally.

Chang, H., & Cartwright, N. (2008). Measurement. In S. Psillos & M. Curd (Eds.), *The Routledge companion to philosophy of science* (pp. 367–375). London: Routledge.

Churchill, G. A. (1979). A paradigm for developing better measures of marketing constructs. *Journal of Marketing Research, 16*(1), 64–73.

Cohen, J. (1990). Things I have learned (so far). *American Psychologist, 45*, 1304–1312.

Cohen, J. (1992). A power primer. *Psychological Bulletin, 112*(1), 155–159.

Cronbach, L. J. (1951). Coefficient alpha and the internal consistency structure of tests. *Psychometrika, 16*(3), 297–334.

Cronbach, L. J., Gleser, G. C., Nanda, H., & Rajaratnam, N. (1972). *The dependability of behavioral measurements: Theory of generalizability for scores and profiles*. New York: Wiley.

De Vaus, D. (2002). *Analyzing social science data*. London: Sage.

Fink, A. (2003). *How to sample in surveys* (2nd ed.). Thousand Oaks, CA: Sage.

Fishbein, M., & Ajzen, I. (1975). *Belief, attitude, intention, and behavior. an introduction to theory and research*. Reading, MA: Addison-Wesley.

Fornell, C., & Larcker, D. F. (1981). Evaluating structural equation models with unobservable variables and measurement error. *Journal of Marketing Research, 18*(1), 39–50.

Groves, R., Fowler, F., Couper, M., Lepkowski, J., Singer, E., & Tourangeau, R. (2009). *Survey methodology* (2nd ed.). Hoboken, NJ: Wiley.

Hair, J., Black, W., Babin, B., & Anderson, R. (2010). *Multivariate data analysis* (7th ed.). Upper Saddle River, NJ: Prentice-Hall.

Homburg, C., Klarmann, M., Reimann, M., & Schilke, O. (2012). What drives key informant accuracy? *Journal of Marketing Research, 49*(4), 594–608.

Horn, J. L., & McArdle, J. (1992). A practical and theoretical guide to measurement invariance in aging research. *Experimental Aging Research, 18*(3), 117–144.

Hunt, S. D. (2012). Explaining empirically successful marketing theories: The inductive realist model, approximate truth, and market orientation. *Academy of Marketing Science Review, 2*(1), 5–18.

Jackson, D. N. (1969). Multimethod factor analysis in the evaluation of convergent and discriminant validity. *Psychological Bulletin, 72*(1), 30–49.

Kardes, F. R. (1996). In defense of experimental consumer psychology. *Journal of Consumer Psychology, 5*(3), 279–296.
Kees, J., Berry, C., Burton, S., & Sheehan, K. (2017). An analysis of data quality: Professional panels, student subject pools, and Amazon's mechanical turk. *Journal of Advertising, 46*(1), 141–155.
Kuß, A., & Eisend, M. (2010). *Marktforschung. Grundlagen der Datenerhebung und Datenanalyse* (3rd ed.). Wiesbaden: Gabler.
Li, R. (Ed.). (2011). *The importance of common metrics for advancing social science theory and research*. Washington, DC: The National Academies Press.
Netemeyer, R. G., Bearden, W. O., & Sharma, S. (2003). *Scaling procedures. Issues and applications*. Thousand Oaks, CA: Sage.
Nunnally, J. C., & Bernstein, I. (1994). *Psychometric theory* (3rd ed.). New York: McGraw-Hill.
Parks, M. R. (2014). Big data in communication research: Its contents and discontents. *Journal of Communication, 64*(2), 355–360.
Peter, J. P. (1979). Reliability: A review of psychometric basics and recent marketing practices. *Journal of Marketing Research, 16*(2), 6–17.
Peterson, R. A. (1994). A meta-analysis of Cronbach's coefficient alpha. *Journal of Consumer Research, 21*(2), 381–391.
Peterson, R. A., & Merunka, D. R. (2014). Convenience samples of college students and research reproducibility. *Journal of Business Research, 67*(5), 1035–1041.
Rentz, J. O. (1987). Generalizability theory: A comprehensive method for assessing and improving the dependability of marketing measures. *Journal of Marketing Research, 24*(1), 19–28.
Rossiter, J. R. (2002). The C-OAR-SE procedure for scale development in marketing. *International Journal of Research in Marketing, 19*(4), 305–335.
Shavelson, R. J., & Webb, N. M. (1991). *Generalizability theory. A primer*. Newbury Park, CA: Sage.
Steenkamp, J.-B. E. M., & Baumgartner, H. (1998). Assessing measurement invariance in cross-national consumer research. *Journal of Consumer Research, 25*(1), 78–90.
Szucs, D., & Ioannidis, J. P. A. (2017). When null hypothesis significance testing is unsuitable for research: A reassessment. *Frontiers in Human Neuroscience, 11*, 390.
Tal, E. (2015). Measurement in science. In E. Zalta (Ed.), *The Stanford encyclopedia of philosophy*. http://plato.stanford.edu
Trout, J. (2000). Measurement. In W. Newton-Smith (Ed.), *A companion to the philosophy of science* (pp. 265–276). Oxford: Blackwell.
Vandenberg, R. J., & Lance, C. E. (2000). A review and synthesis of the measurement invariance literature: Suggestions, practices, and recommendations for organizational science. *Organizational Research Methods, 3*(1), 4–70.
Viswanathan, M. (2005). *Measurement error and research design*. Thousand Oaks, CA: Sage.
Wilson, W. J., & Dumont, R. G. (1968). Rules of correspondence and sociological concepts. *Sociology and Social Research, 52*, 217–227.

Further Reading

Campbell, D. T., & Fiske, D. W. (1959). Convergent and discriminant validation by the multitrait-multimethod matrix. *Psychological Bulletin, 56*(2), 85–105.
Netemeyer, R. G., Bearden, W. O., & Sharma, S. (2003). *Scaling procedures. Issues and applications*. Thousand Oaks, CA: Sage.
Nunnally, J. C., & Bernstein, I. (1994). *Psychometric theory* (3rd ed.). New York: McGraw-Hill.
Peter, J. P. (1979). Reliability: A review of psychometric basics and recent marketing practices. *Journal of Marketing Research, 16*(2), 6–17.
Peter, J. P. (1981). Construct validity: A review of basic issues and marketing practices. *Journal of Marketing Research, 18*(2), 133–145.

Hypotheses and Models for Theory Testing 7

7.1 Hypothesis Testing and Significance Tests

Chapter 2 discusses the close connection between theory and hypothesis. A social science theory contains statements that can or should be verified empirically (Hunt 2010, pp. 188ff.). For such tests, **hypotheses** are central. As explained in Sect. 5.2, these are assumptions about facts (for example, the presence of certain characteristics) or relationships (for example, between attitude and behavior). Now, such general assumptions often cannot be tested at a general level. For example, in the case of the relationship between attitude and behavior, it is often necessary to define specific persons, countries, dates, etc. for an empirical investigation. Hence, one deduces more concrete hypotheses (related to specific situations) from the general theoretical statements (→ deduction, see Sect. 2.5). This is the first step of operationalization, which is followed by the development of measurement instruments, selection of study participants, data collection and data analysis.

In general, one can say that scientific hypotheses are assumptions about facts that go **beyond** the individual case and that can be empirically tested. They represent a link between theory and empiricism. According to Bortz and Döring (2006), the requirements for hypotheses are:

- **Empirical examination**: Scientific hypotheses must relate to real facts that can be empirically investigated.
- **Conditionality**: Scientific hypotheses must be based, at least implicitly, on a meaningful "if-then-proposition" or a "the-more-the-more-proposition". In this sense, assumptions about facts are implicit conditionals. For example, the assumption that, "at least 10% of under-30 year olds have not completed vocational training" can be formulated as, "if a person is under the age of 30, the probability of not having completed vocational training is at least 10%".

M. Eisend, A. Kuss, *Research Methodology in Marketing*,
https://doi.org/10.1007/978-3-030-10794-9_7

- **Generalizability**: Scientific hypotheses must make statements beyond the individual case or a singular event.
- **Falsifiability**: Scientific hypotheses must be (empirically) refutable.

Here are three examples of hypotheses in the field of marketing research:

"The ease of use of a piece of software improves the satisfaction with the software by its users."

"The relationship between attitude and behavior of consumers becomes stronger when the behavior is socially desirable."

"The longer a business relationship lasts, the lower the likelihood that a partner will end the relationship in the near future."

In the context of hypothesis testing using statistical methods, it is important to understand the distinction between alternative and null hypotheses. The **alternative hypothesis** is the statistical formalization of the research question. It is formulated as a statistical assumption that there will be effects, differences, or relationships. The **null hypothesis** contradicts the alternative hypothesis. Research studies usually try to confirm effects (alternative hypothesis). Therefore, the null hypothesis assumes that there are no effects and a hypothesis test attempts to reject the null hypothesis. If it is rejected, one decides to accept the alternative hypothesis. If the null hypothesis cannot be rejected, it will be retained. An alternative hypothesis such as, "satisfied customers are more likely to recommend a product" would be formulated as a null hypothesis thus: "satisfied customers are not more likely to recommend a product". If the hypothesis test shows that this null hypothesis cannot be rejected, then it follows that there is no relationship between customer satisfaction and the likelihood of the customer recommending a product.

A hypothesis is supported if its statement and the corresponding empirical observations are in agreement. However, what does "agreement" mean? The problem of such decisions is illustrated by the following examples:

- We assume (hypothesize) that after at least 10 contacts with brand messages, consumers will actively remember that brand. A study with 200 subjects shows that this was the case for 160 people, but not for the remaining 40 people. Is this result consistent with the assumption?
- We assume (hypothesize) that the intensity of the post-purchase service determines customer satisfaction. In a related study, there is a correlation between these two variables of $r = 0.42$, well below $r = 1.0$ (i.e., a perfect correlation). Is the hypothesis supported?
- We assume (hypothesize) that there is no correlation between the variables "age" and "interest in ecological products", i.e., that the corresponding correlation is at $r = 0$. However, when we investigate the relationship, we find a correlation of $r = 0.08$. Is there a relationship between the two variables?

The questions raised in the first two examples can be clarified easily based on the considerations concerning scientific explanations (see Sect. 2.3.2). Obviously, the first example is not concerned with a regularity which applies to each individual case (→ deductive nomological explanation), but with a *statistical-relevance explanation* that refers to a probability statement (in this case with regard to brand memory). In the second example, we cannot assume that only one variable (post-purchase service) influences another variable (customer satisfaction). Since only one out of a larger number of influencing factors is considered, the relationship between both variables is not perfect or deterministic. Therefore, the resulting correlation is clearly less than 1.0. Rather, in the sense of an explanation based on statistical relevance, we empirically examine whether a substantial correlation (correlation distinctly different from 0) exists between the variables, which would probably be confirmed in the example.

Now to the third and somewhat more complicated example. Here, the question of "**significance**" becomes particularly obvious, that is, the question of whether there is a systematic difference between the expected correlation ($r = 0$) and the measured correlation ($r = 0.08$). The significance or significance level indicates the probability that, in the context of a hypothesis test, the null hypothesis ("there is no systematic relationship") can be erroneously rejected, even though it is actually correct (Type I error, see Sect. 7.3). Therefore, the level of significance is also referred to as the **error probability**. In order to answer the question of significance, we apply *inferential statistics* that serve to make decisions on such questions. In the example case, if one were to take into account the difference between the two values—the desired confidence interval and the sample size with respective distribution assumptions—such a decision could be made. The p-value commonly used for such decisions indicates in this example how large the probability is that a value $r = 0.08$ will be found in the respective sample, if in the population the (actual) value is $r = 0$ (Sawyer and Peter 1983, p. 123). It becomes clear that this is an inductive reasoning, from a relatively small number of cases to an often very large population (for example, the entire population of a country).

A schematic application of statistical methods only for hypothesis tests would be too simple, because all possible errors due to operationalization and measurement would be completely ignored. Such **systematic errors** can be much more serious than sampling errors.

From the point of view of scientific realism (see Chap. 3), one has yet to draw attention to another problem in significance tests. These tests summarize group differences or relationships between variables in a single measure. For example, we may find a positive relationship between variables A and B in 70 or 80% of the subjects studied, but for the remaining individuals, this relationship may be absent or may even be a negative one. However, one would interpret a significantly positive correlation coefficient as having confirmed a suspected positive association. A summary review of several such results would reinforce this effect, giving the impression that these results are quite homogeneous and unambiguous. From the perspective of scientific realism, however, it would make sense to contrast the "empirical successes" with the "empirical failures" (see Sect. 5.3). This aspect is an argument for conducting meta-analyses (see Sect. 9.3).

There are also associations between variables that have no logical relationship, so-called **spurious correlations**. A popular example is the empirically observable relationship between the number of storks and the birth rate in different regions. The reason for this association is obviously a third variable: In the countryside, where there are more storks, there are also more families with many children living there.

W. Lawrence Neuman (2011, p. 413) gives the following assessment of the importance of significance tests:

"Statistical significance tells us only what is likely. It cannot prove anything with absolute certainty. It states that particular outcomes are more or less probable. Statistical significance is not the same as practical, substantive, or theoretical significance. Results can be statistically significant but theoretically meaningless or trivial. For example, two variables can have a statistically significant association due to coincidence with no logical connection between them (e.g., length of fingernails and ability to speak French)."

7.2 Statistical Versus Substantial Significance

The problem discussed above leads to the crucial comparison between **statistical significance** and **substantial significance**. Whether or not a statistical significance occurs depends on various influencing factors. A central influencing factor is the number of cases of an investigation. Figure 7.1 illustrates this relationship using the example of the correlation coefficient. The larger the sample size, the smaller the correlation coefficient, which satisfies the significance criterion of $p < 0.05$, which is frequently used as a critical threshold in marketing research.

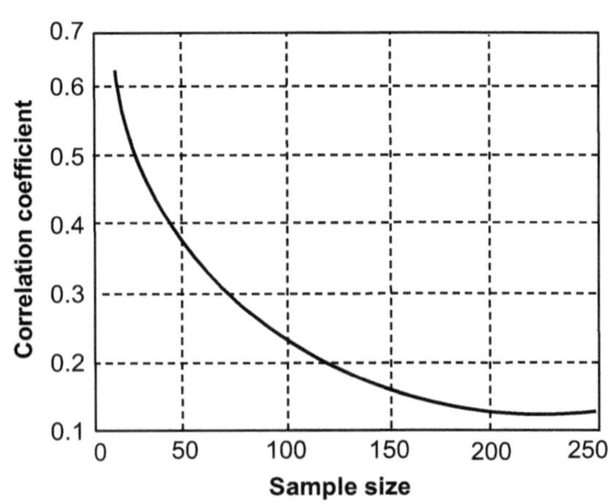

Fig. 7.1 Relationship between sample size and correlation coefficients being significantly different from zero at $p < 0.05$

For very large samples, highly significant results can be found, but they may have only minor theoretical or practical relevance, since the size of the considered effect is very small. In fact, if a sample is large enough, almost every result (e.g., a correlation coefficient) that is only slightly different from zero (or any other comparison value) would be significant. Statistical significance is thus a *necessary but insufficient criterion for a practically or scientifically relevant statement* (that is, for **substantial significance**). For the assessment of the relevance of the hypothesis, effect size is an important criterion that does not dependent on the sample size. We already addressed this problem in Sect. 2.3.2, where an example is provided of the significant difference between substantial and statistical significance.

If statistical significance is given, then the substantial significance can be assessed by the effect size. An **effect size** is "a quantitative reflection of the magnitude of some phenomenon that is used for the purpose of addressing a question of interest" (Kelley and Preacher 2012, p. 140). According to Kelley and Preacher (2012), effect sizes can have different dimensions (e.g., variability, association) and these dimensions are operationalized by different effect size measures or effect size indices. For instance, the dimension of variability can be operationalized in units of variance or standard deviations, the dimension of association in units of correlations. When an effect size measure is applied to data, we obtain an effect size value (e.g., a correlation of 0.25). Marketing research often applies effect sizes that express **associations between variables** or the **strength of relationships** (such as correlations) that indicate how strongly two variables are related, and how large the explanation of the variance of a dependent variable by an independent variable is (Eisend 2015). These effect sizes are commonly applied to describe the relationship between two variables, although some measures of *explained variance* exist that describe the size of an effect that can relate to more than one independent variable and one dependent variable. Effect sizes that measure the extent of the variance explained are central to science, which is above all concerned with explanations. The more science can explain, the better (Aguinis et al. 2011).

The following simple example illustrates the importance of substantial significance and the questionable use of statistical significance in large samples:

A correlation coefficient of 0.03 that measures the relationship between income and happiness is significant at $p < 0.05$ in a sample of 10,000 participants. The result indicates that income is significantly related to happiness (statistical significance). However, the correlation coefficient corresponds to a proportion of explained variance of ca. 0.1 percent. That means 99.9 % of the variation in happiness or income (whatever we apply as dependent variable) remains unexplained, which would be a disappointing figure for scientists who want to explain differences in income or happiness (i.e., its substantial significance).

While significance testing attempts to answer the question of whether there is any difference, effect, or correlation between two variables, the effect size dimension that refers to the strength of a relationship indicates how close the relationship is between two variables. The effect size can be used not only to describe the relationship between two continuous variables (e.g., income and happiness), but also to describe the relationship between two binary variables (e.g., gender and whether someone is a smoker). Although statistical tests for such variables focus on finding out differences or separation (e.g., whether there are more male or female smokers), the test can be understood as one that describes the relationship between gender and smoking. Thus, effect sizes that describe relationships between two variables are appropriate. The effect size shows more meaningful results than a significance test for various reasons:

- Effect sizes *can be compared across different studies and across different types of variables*. This is a common approach in medical science when comparing different studies that examine the effect of different procedures (e.g., type of medication, hospital treatment time, and alternative therapy) in curing the same disease. The higher the explained variance due to a particular procedure (i.e., the closer the association between a procedure and the curing of the disease), the more successful the procedure is.
- Effect size measures such as correlations are often easy to interpret and therefore *more comprehensible to practitioners* than significance tests.
- Effect sizes provide meaningful "*benchmarks*" for comparisons with other study results, between disciplines or even between researchers (Eisend 2015).
- Finally, for each effect size, confidence intervals can also be reported that provide an equivalent to significance tests: if the confidence interval does not contain zero, then the effect is significant (Lipsey and Wilson 2001).

The trend towards "big data" in research, i.e., increasing amounts of data, mainly due to the use of digital technologies, and the issue of the problems of statistical testing on very large samples, has already been addressed in Sect. 6.4. This also explains the increasing importance of effect sizes compared to significance tests. Therefore, a number of scientific journals are placing increasing emphasis on reporting effect sizes while devaluing the importance of significance tests, for example, the "*Strategic Management Journal*" (see Bettis et al. 2016). The journal *Basic and Applied Social Psychology* has even decided not to allow any significance tests (Trafimow and Marks 2015). The current use of significance tests also encourages researchers to engage in dubious practices (e.g., p-hacking, see Chap. 10) to reach results that meet the required significance levels, thus increasing capitalization on chance, biasing the scientific knowledge base and diminishing the probability that results are reproducible (Aguinis et al. 2017).

Another important effect size in marketing research is the **magnitude of an effect**, which provides important information from a substantive and applied perspective. In contrast to the effect size dimension referring to the strength of a relationship, the magnitude of an effect usually applies effect size measures for the *relative change of a dependent variable Y with respect to a relative change of an independent variable X* (e.g., elasticity). This effect size is often of great practical relevance in marketing research because it provides information for an input-output analysis. For example, it can be applied to determine the relative increase in sales of a product due to the relative increase in advertising spending.

The two most important effect size dimensions that are used in marketing research (strength of the relationship and the magnitude of the effect) can thus be distinguished as follows:

- The **strength of a relationship** indicates *how close* the relationship between variables is. Common measures are—amongst others—correlations or proportion of explained variance (see above).
- The **magnitude of an effect**, on the other hand, represents the *extent of change* in a dependent variable due to the change of an independent variable. Common indicators are (unstandardized) regression coefficients.

Both aspects should be additionally illustrated by the example of a linear regression with one dependent and one independent variable (see Fig. 7.2). It shows some (fictitious) measurements and a corresponding regression line. The slope of this line is indicated by a triangle. This slope indicates the magnitude of the effect. Furthermore, the distances of the actually observed values of the dependent variable (y) from the values expected based on the respective x-values and the regression

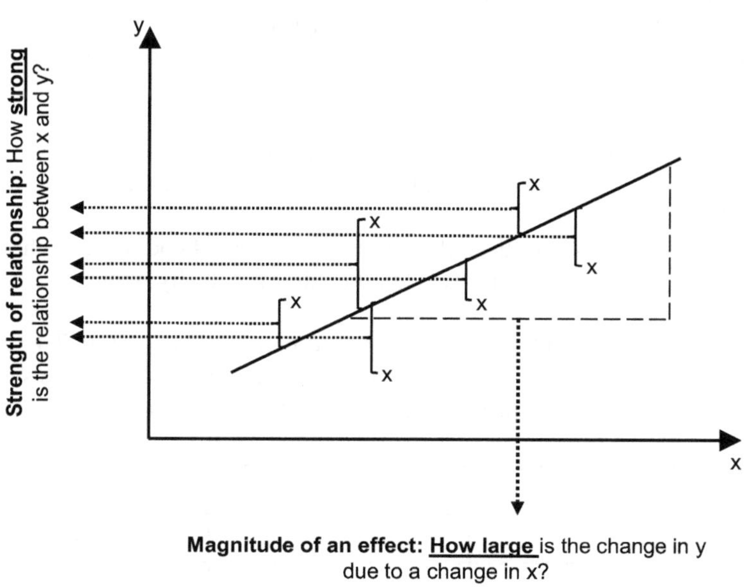

Fig. 7.2 Strength of relationship versus magnitude of an effect

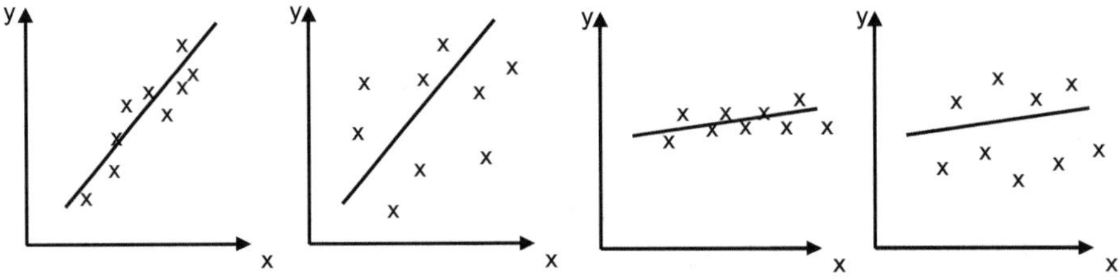

Fig. 7.3 (No) equivalence of strength of relationship and magnitude of an effect

relationship are entered. The smaller these distances are, the stronger the relationship between the variables x and y seems to be.

The strength of a relationship and the magnitude of an effect are therefore not equivalent: a large effect can occur even with little explained variance and thus a weak relationship between two variables. In Fig. 7.3, the difference is illustrated by the relationship between two variables.

7.3 Power of Statistical Tests

The relationship between significance tests, sample sizes and effect sizes is taken into account in the context of "**power analysis**" (Cohen 1988). This analysis addresses the problem of making two mistakes in testing hypotheses:

- One erroneously rejects the null hypothesis. This is a **Type I error**, that is, the mistake of rejecting the null hypothesis, even though it is actually correct. The probability of this is determined by the level of significance or the error probability.
- One erroneously assumes the null hypothesis. This is a **Type II error**, the mistake of accepting the null hypothesis, even though it is actually wrong.

The four possible results of a significance test are depicted in Fig. 7.4.

The smaller the α-error in a study, the less frequently the null hypothesis is falsely rejected. This increases the probability of mistakenly accepting the null hypothesis

	H_0 is true	**H_0 is false**
H_0 not rejected	Correct decision	Type II error (β-error)
H_0 rejected	Type I error (α-error)	Correct decision

Fig. 7.4 Results and errors of hypothesis testing

and rejecting the alternative hypothesis (β-error). However, the size of the α-error does not directly deduce the size of the β-error and vice versa. The two types of errors are determined in different ways. The size of the α-error depends on the significance level.

The size (1-β) is also referred as **power** (Cohen 1988). The power of a test (that is, the likelihood that testing a null hypothesis leads to rejection of the null hypothesis if the alternative hypothesis is correct) is influenced by three factors (next to the variance):

- α-significance level: the smaller α, the lower the probability of choosing the alternative hypothesis falsely (Type I error);
- Sample size: the larger the sample size, the greater the probability of deciding in favor of the alternative hypothesis (ceteris paribus);
- Effect size: the larger the explained variance and the strength of a relationship, the greater the power of the test and thus the probability of deciding against the null hypothesis and in favor of the alternative hypothesis.

In summary, at a given significance level (e.g., $\alpha = 0.05$) larger effect sizes tend to become more likely significant than smaller effect sizes and larger samples have higher test sensitivity than small samples, and thus are more likely to produce significant results.

Although there are no formal **standards for power levels** (also referred to as π (pi)), a value of $\pi = 0.80$ is usually used, that is, a four-to-one probability between β-error and α-Error (Ellis 2010). If the test is designed in such a way that it should not produce any β errors, then a lower standard can be applied. This is often the case in medical research, where it is better to assume that one has an indication of a disease, even if the patient is healthy, than to assume that a patient is healthy, but in reality, is suffering from a disease.

Power analysis is important for the *interpretation of test results*, because the power indicates the probability of correctly rejecting the null hypothesis. It is, as already explained, dependent on the chosen significance level, the effect size, and the sample size. This attests to the central idea that a hypothesis can be rejected for various reasons. A hypothesis may be rejected because the effect is too small, which is easy to understand and desirable from a scientific point of view. However, a hypothesis can also be rejected because the sample is not large enough or the significance level is too small, that is, it was chosen as being too strict. With an increase in the sample size or a "more generous" level of significance, the hypothesis could possibly be accepted based on the same data.

How to choose the *right significance level*? The social sciences have established a significance level of 5%, proposed by Ronald Fisher (1925, p. 43). This limit means that, on average, one in twenty studies in which the null hypothesis is correct (e.g., age is not related to happiness) is found to be false (e.g., age is related to happiness). Sometimes results are accepted even at a lower significance level of <0.1. Which levels of significance are accepted also depends on the degree of innovation of a study: scientists tend to apply less stringent criteria to completely new and innovative results, and possibly consider marginally significant results to be relevant, than to results that relate to an already established hypothesis. Depending on the object under investigation, a Type I error may be less serious than a Type II error, as indicated in the above example in medical science, where one is more likely to accept a disease, even if the patient is healthy, than to assume that a patient is healthy, when she or he is actually ill.

The relationship between level of significance, effect size, and sample size also makes it possible to *determine the sample size* for a known or expected effect size that is necessary so that the effect at a given level of significance with a desired power is actually significant. It can already be seen in Fig. 7.1 that large effect sizes require smaller samples in order to reach the specified significance level and vice versa. In addition, if the power level is high, the sample size needed to reach the significance level continues to increase, especially with small effect sizes.

7.4 A Priori Hypotheses Versus Post Hoc "Hypotheses"

The usual applications of statistical tests are based on a procedure in which one hypothesis or a few specific hypotheses are formulated, then appropriate data are collected and suitable statistical tests are applied. Examples include studies on the efficacy of drugs (new drug vs. placebos) or testing previously theoretically well-founded hypotheses (such as the relationship between x and y). Figure 7.5 illustrates this "classical" approach to testing (a few) **hypotheses that have been formed a priori**. It shows the path from a few theoretically well-founded hypotheses to data

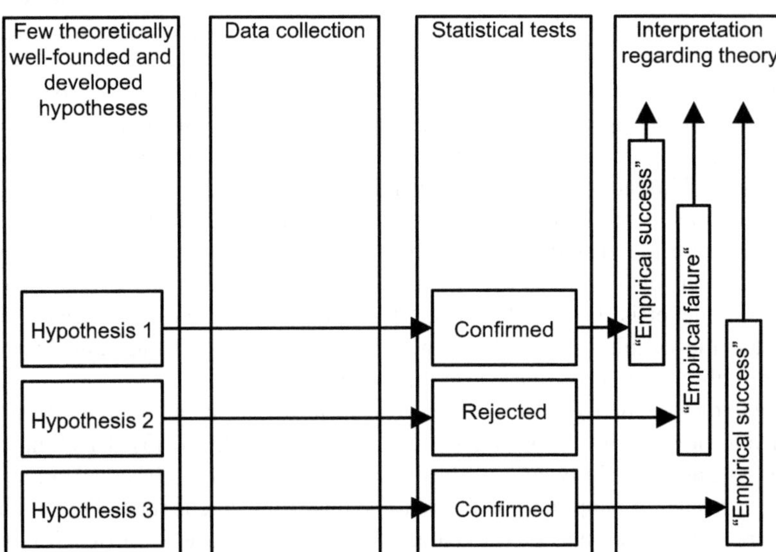

Fig. 7.5 Procedure of testing a priori hypothesis

collection and statistical tests that confirm or reject the hypotheses and to their interpretation as "empirical successes and failures" (see Sect. 5.3). This procedure also corresponds to the hypothetico-deductive method described in Sect. 5.2.

In many marketing research studies, however, data collection is not limited to very few selected variables. It is more common to include a larger number of variables. For example, survey questionnaires usually include a two-digit number of questions with a corresponding number of variables. Under these conditions, researchers may choose from a variety of possible (and easy to compute) correlations those that appear to be "significant", and hypothesize the relationships later, because it is easier to publish significant results than non-significant ones (see Sect. 10.2.4). With the goal of increased publication opportunities (which is, under today's publication pressure, partially understandable), theories and hypotheses are adapted to already existing results; that is, **post hoc hypotheses** are formulated. These are not real hypotheses (see Sect. 7.1) because, given already existing results, one cannot speak of assumptions and falsifiability is not possible. The problem is not unknown in the literature (and probably also in research practice): Peter (1991, p. 544) speaks of "fishing through a correlation matrix"; Kerr (1988) speaks of "**HARK**ing: **H**ypothesizing **A**fter the **R**esults are **K**nown"; Leung (2011) discusses "Presenting Post Hoc Hypotheses as A Priori ...". Already some fifty years ago, Selvin and Stuart (1966) referred to such an approach as "data dredging". The extent of the problem in research practice is difficult to know, because in such cases, the authors avoid disclosure and readers of articles based on HARKing find few clues. Banks et al. (2016) reported in a study that about 50% of respondents to a survey in management research said they had "presented a post hoc hypothesis as if it were developed a priori" (p. 10). The problem concerns research ethics (see Sect. 10.2.4) and can lead to grossly misleading results. The reasons are briefly outlined below.

The starting points of the considerations are the following real-life experiences:

- Researchers are anxious to find significant results because their chances of publication are much greater than for those of non-significant ones.
- For a larger number of potential associations of variables, by chance, some seemingly "significant" relationships arise, even if no such relationships actually exist. Even if one correlates numerous variables, which were generated by random numbers, for which there can be no systematic relationship, a few correlation coefficients would be "significantly" different from zero and misleadingly indicate that there are real relationships (Kruskal 1968).

The problem is illustrated by a very simple example. Figure 7.6 shows a (hypothetical) correlation matrix for the variables A to H, which are measured in a reasonably large sample. In the corresponding population, there is no correlation between any of these variables, so that the corresponding correlation coefficients are (or should be) 0. Accordingly, in the correlation matrix for the (sample) data, in the main diagonal are the "1" values and in the other fields are values which are very close to 0 (ideally the value 0). However, it may well be that through sampling, some cases were sampled that led by chance to some correlation coefficients that are

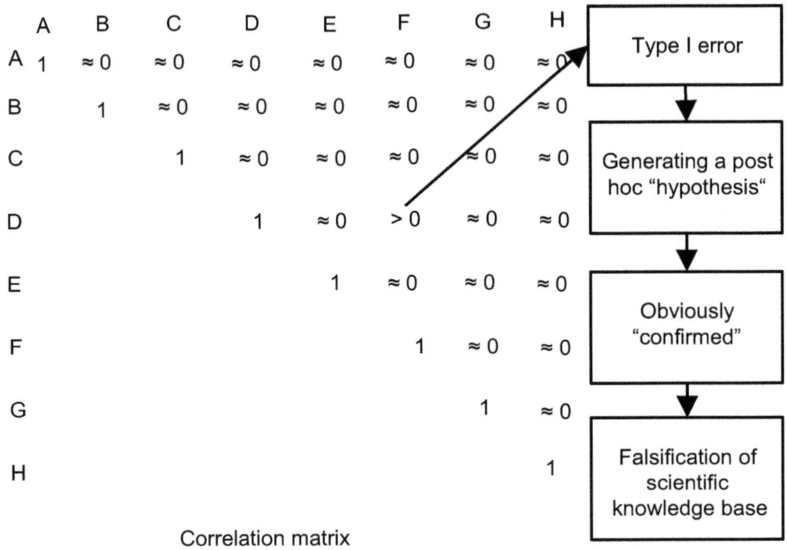

Fig. 7.6 Procedure of generating post hoc "hypotheses"

clearly greater than 0 (thus apparently "significant"). In the example this is entered for the variable combination D and F, which is marked with "> 0". This would correspond to a Type I error (see above), because the correct null hypothesis would be rejected. If, following from this result, a (post hoc) "hypothesis" is proposed, then its (apparent) confirmation would be unavoidable because the corresponding result is already known. Kerr (1988, p. 205) uses the ironic phrase "HARKing can translate type I errors into theory". Furthermore, for hypotheses that have been formed subsequently, the requirement (see Sect. 7.1) that hypotheses can be rejected by the investigation is violated. The interpretation of such a random result as a statistical confirmation of a previously theoretically developed hypothesis would be misleading in regard to the relevant scientific knowledge.

An example of a problematic use of significance testing is a study concerning personality traits and consumer behavior that appeared in the early years of the highly respected *Journal of Marketing Research*. The study was about relationships between personality traits (e.g., aggression) and consumer behavior. For this purpose, the relationships between three personality variables and 15 characteristics of consumer behavior (product use, brand preferences in different product groups) were examined (with a relatively weak database) by means of Chi^2 tests. In these 45 tests, there were seven (apparently) significant relationships. For example, an association has

(continued)

emerged between aggressiveness and the preference for wet or electric shaving, a connection that may not be theoretically compelling. It is questionable which proportion of the seven "significant" results has a real basis or came about by chance.

To distinguish from such an approach is the test of so-called **implicit hypotheses**. These are hypotheses that do not belong to the core of the theoretical question and are not necessarily fixed a priori (e.g., fixed in writing). However, for these hypotheses, the researcher collects corresponding additional data due to his or her experience and theoretical training, which suggests that there might still be interesting or relevant relationships (e.g., as a control variable). This would lead to a rather small number of additional hypotheses for which the statistical problem outlined above appears only to a limited extent. One may well assume that the "temptation" to HARKing is greatest when large (many variables) data sets, that are not self-collected, are used. On the other hand, in the case of one's own data collection, one usually deals with a restricted number of variables that were considered meaningful and important at the *beginning* of the investigation and then collected. The least likely is the problem in experimental studies (see Sect. 8.3), which is confined to a small number of carefully established variables.

It goes without saying that the description and documentation of particularly interesting results, which are not based on previously developed hypotheses, are of course possible, but not with the claim of statistical confirmation. If post hoc hypotheses are to be verified empirically / statistically, then another data set is required that is independent from the data from which this hypothesis was created. Furthermore, the interpretation of data without a priori hypotheses can make sense when applying an *inductive approach*. In any case, researchers need to be transparent about what they do. The problem of HARKing mainly refers to a lack of transparency, that is, when researchers present post hoc hypotheses as a priori hypotheses without acknowledging having done so (Kerr 1988).

7.5 Modeling with Regression Analysis

In the context of testing theories, some areas of marketing research use mathematical models for the presentation and solution of problems. Mathematically formalized modeling is also simply referred to as **modeling**. This approach is based above all on econometrics, a branch of economics that combines economic theory, empirical data and statistical methods. The central task of econometrics is the derivation of econometric models from economic theories and their numerical concretization. With the help of econometrics and modeling, interesting relationships in economics can be quantified (e.g., percentage change of the savings rate with percentage change of the interest rate), thus hypotheses and whole models can be empirically tested and these empirically validated models can be used for *forecasts* or *simulation* (e.g., economic growth will change as inflation rates change).

In marketing research, in addition to the applications of econometrics, *optimization questions* are often at the forefront of modeling. Shugan (2002) distinguishes between two different definitions of mathematical models, one being the mathematical optimization of variables and the other mathematical mapping with the purpose of solving research questions. In the former view, it is often sufficient to show that a particular solution is optimal, for example, what is the optimal ratio between advertising spending and personal selling? It is often about optimizing resource allocations. In addition to such models, which are oriented towards solving practical problems, they also serve to develop a theoretical understanding of marketing problems by varying assumptions and determining the resulting changes in dependent variables. Often the second approach does not involve a systematic empirical review of the model assumptions, but a fair presentation of the adequacy and successful application of such models based on selected cases is very common.

Parameterization and validation, in the context of modeling, use methods that are based on classical **regression analysis**. Regression analysis is a statistical method that attempts to explain the change in a *dependent variable* by changes in a set of so-called *explanatory* or *independent variables* by quantifying a single equation. A regression can determine whether there is a quantitative relationship between the independent variables and the dependent variable. However, the result of a regression analysis alone cannot show causality even when statistical significance is given, since a statistical relationship never implies causality (for causality and the special requirements for the appropriate study design, see Chap. 8). Nevertheless, regression analysis and other econometric techniques are used to determine relationships between variables, which are often interpreted as cause-and-effect relationships. In order for the empirical regression analysis to be done, strict assumptions must be fulfilled.

In the simplest case, a regression model $Y = \beta_0 + \beta_1 X_1 + \ldots + \beta_k X_k + \varepsilon$ describes an endogenous variable Y by a linear relationship to one or more other (exogenous) variables $X_1, \ldots, X_k$. The model explains the **endogenous** variable, while an **exogenous** variable is explained, not by the model, but by variables outside the model. For instance, if the regression model tries to explain the degree of happiness by variables such as age and income, happiness is an endogenous variable in the model, while age and income are exogenous variables. Of course, income can be explained by other variables such as education, but since they are not included in the model, income is considered an exogenous variable in that particular model. Since in practice there will be no exact relationship between the empirically observed variables, so that the exogenous variables could fully explain the endogenous variable, an error term in the equation records all variables that besides $X_1, \ldots, X_k$ also have an influence on Y. After specifying a particular model, the model parameters $\beta_0, \ldots, \beta_k$ are estimated. On this basis, forecasts can be made for the values of Y for assumed values of $X_1, \ldots, X_k$.

Regarding the usual results of a regression analysis, we find results that stand for significance and measures for effect sizes in terms of the strength of a relationship and magnitude of the effect (see Sect. 7.2):

- **Strength of a relationship/explained variance**: The corresponding measure R^2 *(coefficient of determination)* shows what proportion of the variance of the dependent variable is explained by all the independent variables.
- **Magnitude of an effect**: The *unstandardized regression coefficients* β_0, ..., β_k indicate how much a change of the respective independent variable affects the dependent variable, that is, by what extent the dependent variable changes, if the independent variable changes by a certain extent. This value depends on the scaling of the variable. Thus, for example, the magnitude of the effect that measures the relationship between advertising spending and sales (units sold) depends on whether we measure the spending in US dollars, euros, or Swiss francs. If these coefficients are specified as elasticities, that is, the ratio of the percentage change in one variable (e.g., sales) to the percentage change in another variable (e.g., advertising spending), the scaling problem is eliminated.
- **Significance of the regression model:** Tests are used to check whether the proportion of explained variance (R^2) is significantly different from 0, that is, whether the model makes (at least a small) contribution to the explanation of the dependent variable (see also Sect. 2.3.2).
- **Significance of the regression coefficients**: With t-tests, we check whether the different regression coefficients β are significantly different from 0. Otherwise—at $\beta = 0$—a change in the respective independent variable would have no systematic effect on the dependent variable.

The standard method for estimating the parameters in linear regression models is the **Ordinary Least Squares (OLS)** estimation. In order to be able to apply it without problems, a number of assumptions have to be fulfilled, which also have important substantive implications with regard to theory testing (Allison 1999; Hair et al. 2010):

- The regression model must have *parametric linearity* (i.e., the relationship of the variable must follow a linear function) and not all observations of an X variable may be the same (i.e., they must vary), otherwise no estimation is possible.
- The conditional expected value of the error term must be zero, which implies a covariance between the X variables and the error term of zero. This assumption of the *exogeneity* of X_1, ..., X_k is important, because only in this case are ceteris-paribus statements, such as "a change of X_1 by one unit leads to a change of Y by β_1 units," possible. For instance, the influence of advertising spending on sales can lead to endogeneity problems, because advertising spending decisions often depend on sales in prior periods and are therefore not exogenous to the model. A statement such as "a change of 10% in advertising spending leads to a change of 3% in sales" would be wrong, since the change in sales also depends on the sales of the prior period, as does the change in advertising spending.

- The conditional variance of the error term must be constant. A famous example for a violation of this condition is the relationship between income and spending on certain consumption activities, such as food. At a low income, consumers spend a certain constant amount on food, as they cannot afford more. With increasing income, consumers display a greater variation in spending on food, as they sometimes buy inexpensive food but at other times enjoy expensive meals. As a result, the error term variance would increase with the increase in the independent variable.
- The conditional error term covariance must be equal to zero, which means that the data point deviation from the regression line does not show any pattern along the independent variable. This is often violated in *time series data*, where the independent variable is time. Most data points show a particular pattern over time, for example an economic cycle, and a data point is not independent of the preceding data point (e.g., if the economy shows high economic growth in one year, it probably shows relatively high economic growth in the following year, too). As a result, the error terms show a co-variation pattern.
- There must be no perfect correlation between the explanatory variables, since in this so-called perfect *multicollinearity* an OLS estimation is impossible. In addition, imperfect multicollinearity, characterized by high correlations between explanatory variables, is problematic, because in this case OLS cannot precisely distinguish between the influences of the individual variables and cannot provide accurate parameter estimates.
- The error terms should be normally distributed.

One can use a number of statistical tests to obtain evidence for a *violation of these assumptions*. When violations are identified, the model specification can be revised, robust procedures can be used, or alternative estimation techniques (such as instrumental variables) can be used, depending on the nature of the problem. If the theory already suggests that assumptions of the classical regression model are not realistic (e.g., a correlation of the error terms occurs regularly with time series data), alternative estimation methods are usually used right from the start. The following is a brief illustration of how to deal with the violation of the respective assumptions (for more detail, see Allison 1999 or Gujarati 2003).

- If the assumption of the **parameter linearity** is not met, a parameter-linear form can be produced by variable or model transformation (for example by log transformation). Meanwhile, there are also estimation methods for non-linear relationships (non-linear least squares).
- **Endogeneity** can be detected with the Hausman test. To solve the endogeneity problem, one can introduce an instrumental variable (IV estimation). This requires so-called instrumental variables that are highly correlated with the endogenous explanatory variables (instrument relevance) and at the same time are not correlated with the error term (instrument exogeneity). Given the proper quality of the IV estimator, consistent parameter estimates are achieved. The

quality of the instruments can be checked by regressing the endogenous explanatory variable on all instruments, including the exogenous variables.

- Whether or not the problem of **heteroscedasticity** occurs (i.e., not a constant conditional variance of the error term) can also be tested, using either the Breusch-Pagan or White test. In the case of heteroscedasticity, robust error terms can be used instead of the standard error terms, which the OLS wrongly estimates. Alternatively, the use of WLS (Weighted Least Squares) is conceivable in large samples.
- In time series regressions (i.e., data are collected repeatedly at different points in time), one often faces the problem of error term **autocorrelation**, which is detected by various tests (Durbin-Watson test, Breusch-Godfrey test). Again, one has the opportunity to use autocorrelation robust standard errors or to estimate a GLS (Generalized Least Squares) model. This procedure provides correct standard errors, and more efficient estimates of the model parameters, if the autocorrelation structure used for the model transformation is correctly recognized and implemented in the new model.
- Perfect **multicollinearity** is unlikely to occur in social science research, but high multicollinearity can occur. High multicollinearity is often recognized by high pairwise correlations between the independent variables and high coefficients of determination in models, in which one exogenous variable is explained by all other exogenous variables. The Variance Inflation Factor (VIF), or Tolerance, measures multicollinearity. High multicollinearity is avoided by excluding variables from the regression model or by grouping variables into factors or indices.
- The **assumption of the normal distribution of the error term** is usually not subject to intensive tests in practice. Due to sufficiently large samples, a normal distribution of the estimated parameters can be assumed due to the central limit theorem.

7.6 Structural Equation Models

Structural equation models work as a test of networks of hypotheses or larger parts of theories. The alternative designation of **causal models** is somewhat problematic, similar to regression analysis, because the application is often based on cross-sectional data that do not allow a proof of causality. This is outlined in Chap. 8: "The ability to make a causal inference between two variables is a function of one's research design, not the statistical technique used to analyze the data that are yielded by that research design." (Jaccard and Becker 2002, p. 248). Finally yet importantly, it is difficult to exclude alternative explanations for a common variation of causes and effects (see Sect. 8.1).

The basic idea of Structural Equation Models (SEM) is that, based on the variances and covariances of indicators (observable variables) found in a dataset, conclusions are drawn with respect to *relationships between complex constructs* (latent variables). The characteristic features of structural equation models can be seen in the fact that a larger number of interconnected relationships is analyzed, and at the same time not directly observed concepts could be included in these relationships, whereby measurement errors can be explicitly taken into account.

The following is an illustration of the simultaneous analysis of multiple relationships, whereby possible measurement errors are not taken into account. The underlying model is the Technology Acceptance Model (TAM) of Davis et al. (1989), widely applied in technology use research, which explains the acceptance and use of (computer-based) technologies. A simplified model is depicted in Fig. 7.7. It assumes that the intention to use a technology depends on the perception of the usefulness of this technology (H1) and the ease of use (H2). The ease of use also influences the perceived usefulness (H3). Intention to use increases the actual use (H4). It can be seen that in this model several hypotheses or a part of a theory are *simultaneously* considered and (later) tested.

Such a model is called a structural model. It describes *relationships between the latent variables* (concepts). These variables cannot be observed directly, but can be estimated using appropriate measurement models. The next step is the development and application of these measurement models (similar to scale development, see Sect. 6.2), so that the parameters of the model can be estimated. For this purpose,

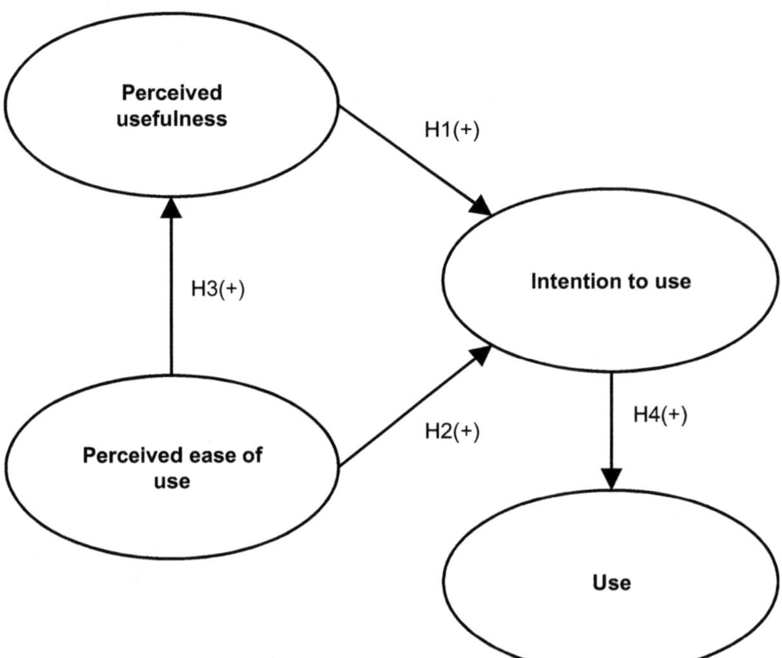

Fig. 7.7 Example of a structural model (simplified Technology Acceptance Model by Davis et al. 1989)

different **indicators** are used in the present example for the different latent variables. For example, the perceived usefulness of a technology can be measured with the following indicators. Respondents indicate the extent to which they agree with these statements on a scale ranging from 1 ("totally disagree") to 7 ("fully agree") for the endpoints:

- Productivity: "Using this technology facilitates productivity."
- Effectiveness: "Using this technology facilitates effectiveness."
- Performance: "Using this technology improves performance."

Accordingly, all latent variables are measured by appropriate indicators (all are *manifest variables*). The (simplified) representation of the structural model with the corresponding measurement models is depicted in Fig. 7.8.

Measurement errors are considered in such models in two ways: Each indicator (e.g., "productivity" or "effectiveness") is associated with a measurement error that is unobservable. The idea behind it is analogous to a regression model. In that, the latent variable explains the indicator, with the measurement error added like an error term in the regression analysis, because the explanation is not complete. Similarly, endogenous latent constructs (that is, variables that are explained by other constructs in the model, e.g., "intention to use") are each assigned a measurement error that captures the unexplained variance next to the explained variance by the constructs influencing them (e.g., "perceived ease of use").

The measurement error design is due to indicators that are called **reflective indicators**, that is, indicators that are caused by the latent variable. Accordingly, the arrows in the model are directed so that the latent variable and the measurement error explain an indicator. Thus, it is assumed that the latent variable (e.g., "perceived usefulness") causes the different expressions of the indicators ("productivity",

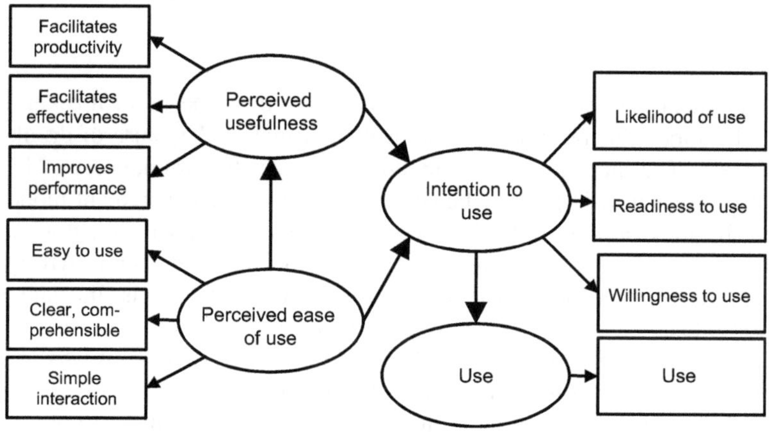

Fig. 7.8 Example of a structural and measurement model (simplified Technology Acceptance Model by Davis et al. 1989—illustration without measurement errors)

"effectiveness", "performance"). This is a perfectly plausible assumption in many social psychological phenomena where it is assumed that an observation (e.g., a verbal opinion) can be explained by an underlying concept: for example, a statement such as, "I like the Apple brand" is "caused" by the attitude to the Apple brand. However, there are also constructs in which the latent variable is explained or caused by the indicators. These indicators are referred to as **formative indicators** (for more detail on reflective vs. formative indicators see Burke et al. 2003).

The difference between formative and reflective indicators can be clearly illustrated by the example of drunkenness and fitness to drive (see Ringle et al. 2006, p. 83). The model is simplified and shows no measurement errors.

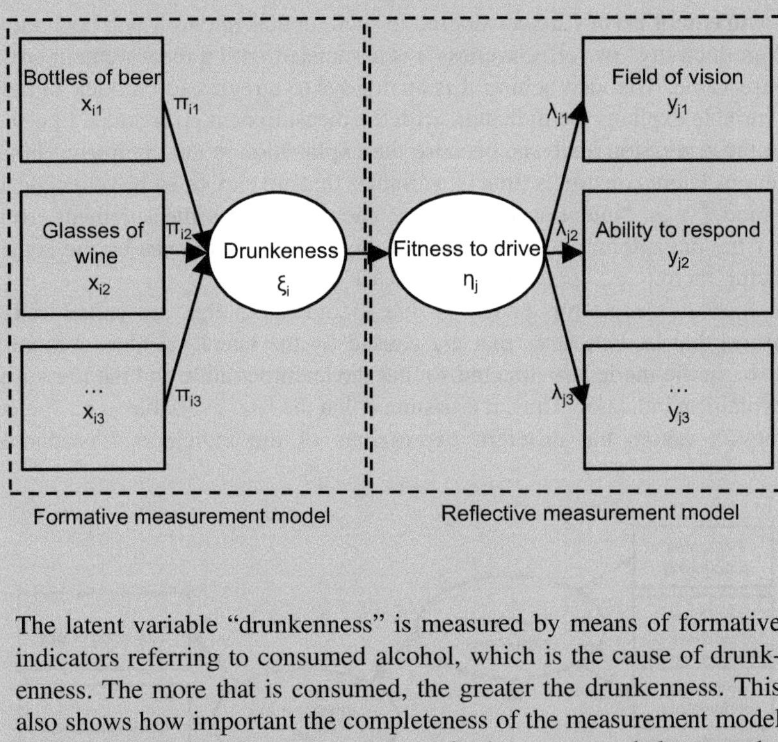

The latent variable "drunkenness" is measured by means of formative indicators referring to consumed alcohol, which is the cause of drunkenness. The more that is consumed, the greater the drunkenness. This also shows how important the completeness of the measurement model is. If, for example, only the amount of wine consumed, but not the amount of beer consumed, is measured, the measurement is wrong. Unlike formative ones, for reflective measurement models, the latent variable is the origin of changes in the indicator values. As a result, all the indicators associated with a latent variable are highly correlated, so that the elimination of a single reflective indicator is usually not a problem. In the example, the fitness to drive has an influence both on the size of the field of vision and on the ability to respond.

Structural equation models, in particular the measurement models contained therein, are also often used today to test the **convergent and discriminant validity of measurements** of constructs (see Sect. 6.3.3). On the one hand, the correspondence of several indicators for the measurement of the same construct (→ convergent validity) is tested, and on the other hand, the discriminability of several constructs (→ discriminant validity) are examined (Fornell and Larcker 1981; Hair et al. 2010).

Estimating the parameters of such models requires complex and sophisticated procedures for which appropriate software is available, although, of course, this does not obviate the need for a thorough understanding of the methods to ensure a meaningful application. Software is distinguished into covariance-based techniques (e.g., LISREL / AMOS / MPlus) and variance-based methods (PLS). The result of such an estimation shows whether or not the theoretically suspected relationships between the different variables are confirmed and how strong these relationships are. For such results, so-called fit indices are used to assess the extent to which the theoretical model complies with the data collected. These methodically challenging questions are widely discussed in the literature. For (relatively) easy-to-understand presentations, see Hair et al. (2010).

References

Aguinis, H., Dalton, D. R., Bosco, F. A., Pierce, C. A., & Dalton, C. M. (2011). Meta-analytic choices and judgment calls: Implications for theory building and testing, obtained effect sizes, and scholarly impact. *Journal of Management, 37*(1), 5–38.

Aguinis, H., Cascio, W. F., & Ramani, R. S. (2017). Science's reproducibility and replicability crisis: International business is not immune. *Journal of International Business Studies, 48*(6), 653–663.

Allison, P. D. (1999). *Multiple regression: A primer*. Thousand Oaks, CA: Pine Forge.

Banks, G., O'Boyle, E., Pollack, J., White, C., Batchelor, J., Whelpley, C., et al. (2016). Questions about questionable research practices in the field of management: A guest commentary. *Journal of Management, 42*(1), 5–20.

Bettis, R. A., Ehtiraj, S., Gambardella, A., Helfat, C., & Mitchell, W. (2016). Creating repeatable cumulative knowledge in strategic management. *Strategic Management Journal, 37*(2), 257–261.

Bortz, J., & Döring, N. (2006). *Forschungsmethoden und Evaluation* (4th ed.). Berlin: Springer.

Burke, C. B., MacKenzie, S. B., & Podsakoff, P. M. (2003). A critical review of construct indicators and measurement model misspecification in marketing and consumer research. *Journal of Consumer Research, 30*(2), 199–218.

Cohen, J. (1988). *Statistical power analysis for the behavioral sciences* (2nd ed.). New York: Academic.

Davis, F. D., Bagozzi, R. P., & Warshaw, P. R. (1989). User acceptance of computer technology: A comparison of two theoretical models. *Management Science, 35*(8), 982–1003.

Eisend, M. (2015). Have we progressed marketing knowledge? A meta-meta-analysis of effect sizes in marketing research. *Journal of Marketing, 79*(3), 23–40.

Ellis, P. D. (2010). *The essential guide to effect sizes: An introduction to statistical power, meta-analysis and the interpretation of research results*. Cambridge: Cambridge University Press.

Fisher, R. A. (1925). *Statistical methods for research workers*. Edingburgh: Oliver & Boyd.

Fornell, C., & Larcker, D. F. (1981). Evaluating structural equation models with unobservable variables and measurement error. *Journal of Marketing Research, 18*(2), 39–50.
Gujarati, D. N. (2003). *Basic econometrics* (4th ed.). Boston: McGraw-Hill.
Hair, J., Black, W., Babin, B., & Anderson, R. (2010). *Multivariate data analysis* (7th ed.). Upper Saddle River, NJ: Prentice-Hall.
Hunt, S. D. (2010). *Marketing theory—Foundations, controversy, strategy, resource—Advantage theory*. Armonk, NY: Sharpe.
Jaccard, J., & Becker, M. (2002). *Statistics for the behavioral sciences* (4th ed.). Belmont, CA: Wadsworth.
Kelley, K., & Preacher, K. J. (2012). On effect size. *Psychological Methods, 17*(2), 137–152.
Kerr, N. (1988). HARKing: Hypothesizing after the results are known. *Personality and Social Psychology Review, 2*(3), 196–217.
Kruskal, W. (1968). Tests of statistical significance. In D. Sills (Ed.), *International encyclopedia of the social sciences* (pp. 238–250). New York: Macmillan and Free Press.
Leung, K. (2011). Presenting post hoc hypotheses as a priori: Ethical and theoretical issues. *Management and Organization Review, 7*(3), 471–479.
Lipsey, M. W., & Wilson, D. T. (2001). *Practical meta-analysis*. Thousand Oaks, CA: Sage.
Neuman, W. L. (2011). *Social research methods—Qualitative and quantitative approaches* (6th ed.). Boston: Pearson.
Peter, J. (1991). Philosophical tensions in consumer inquiry. In T. Robertson & H. Kassarjian (Eds.), *Handbook of consumer behavior* (pp. 533–547). Englewood Cliffs, NJ: Prentice-Hall.
Ringle, C., Boysen, N., Wende, S., & Will, A. (2006). Messung von Kausalmodellen mit dem Partial-Least-Squares-Verfahren. *Wirtschaftswissenschaftliches Studium, 35*(1), 81–87.
Sawyer, A. G., & Peter, J. P. (1983). The significance of statistical significance tests in marketing research. *Journal of Marketing Research, 20*(2), 122–133.
Selvin, H., & Stuart, A. (1966). Data-dredging procedures in survey analysis. *The American Statistician, 20*(3), 20–23.
Shugan, S. (2002). Marketing science, models, monopoly models, and why we need them. *Marketing Science, 21*(3), 223–228.
Trafimow, D., & Marks, M. (2015). Editorial. *Basic and Applied Social Psychology, 37*(1), 1–2.

Further Reading

Ellis, P. D. (2010). *The essential guide to effect sizes: An introduction to statistical power, meta-analysis and the interpretation of research results*. Cambridge: Cambridge University Press.
Hair, J., Black, W., Babin, B., & Anderson, R. (2010). *Multivariate data analysis* (7th ed.). Upper Saddle River, NJ: Pearson.
Jaccard, J., & Becker, M. (2002). *Statistics for the behavioral sciences* (4th ed.). Belmont, CA: Wadsworth.

Testing Causal Relationships 8

8.1 Essence and Relevance of Causality

People have a fundamental need to "get to the bottom of things", that is, to understand the reason behind, for example, the reasons for the course of the stars or the ways to live a happy life or the causes of economic growth. People are looking for explanations. Godfrey-Smith (2003, p. 194) puts it concisely: "To explain something is to describe what caused it". It is therefore not surprising that questions of causality, the *search for causes and effects*, have for a long time occupied people, in particular scientists. There are differing views and comprehensive discussions on the nature and characterization of causality in the philosophy of science (see, for example, Godfrey-Smith 2003, pp. 194ff.).

In Chap. 7, the focus was on the testing of hypotheses. In a scientific context, the test of **relationships** between variables is of particular interest. This chapter deals with a special kind of relationship, so called **causal relations**, which have particular significance and—because of that—place particular demands on the nature of the relationships between variables. The first section deals with the essential features of **causality**, then types of causal relationships are outlined. Other parts of this chapter deal mainly with basic ideas about conducting experiments, which is the most common method for the study of causal relationships.

The philosophical literature has dealt with the question, "What is causality?" for nearly 400 years. Of course, this textbook does not try to discuss and understand this stream of literature in its entirety. Introductions and summaries are offered, amongst others, by Humphreys (2000), Mumford and Anjum (2013) and Psillos (2002). Even those who cannot or will not understand the details of this discussion will be easily able to assess the *relevance of causality* through a few examples. The following examples from different areas of business, society and science/technology show "that the concept of causality is a heuristic that helps us to think about our environment, organize our thoughts, predict future events, and even change future events" (Jaccard and Jacoby 2010, p. 140). Based on these examples (see Mumford and

M. Eisend, A. Kuss, *Research Methodology in Marketing*,
https://doi.org/10.1007/978-3-030-10794-9_8

Anjum 2013, p. 1), more general features of causality will be characterized in the following sections.

1. **Society**: Individual behaviors have consequences, for example, careless parenting is considered a possible cause of children's poor academic performance. If there was not a causal link, one could not speak of (co-)responsibility of the parents.
2. **Law**: Human behavior (for example, in traffic) can cause physical or material damage to other people. Without a causal relationship (behavior → damage), there could be no evidence of guilt or claims.
3. **Technology**: In the case of accidents, technical defects, etc., one typically looks for the causes (causes of accidents, etc.), on the one hand to clarify the responsibility and to derive a claim settlement from it. On the other hand, one wants to learn from it and reduce or eliminate such risks in the future. This often requires the analysis of a *causal chain*, i.e., the individual steps between a cause and the resulting consequences or effects (see Sect. 8.2). Thus, the collapse of a bridge (in a nonprofessional's conception) could have come about through the following causal chain: steel reinforcement of the concrete bridge poorly protected against moisture → rapid rusting of load-bearing parts → instability of the bridge → collapse.
4. **Medicine**: Medical research and practice looks for the corresponding causes of disease symptoms in order to develop a therapy (e.g., high blood pressure increases the risk of infarction).
5. **Economics**: Almost daily, the media report and analyze more or less well founded or speculative causes of current macroeconomic developments, for example, "Growing domestic demand causes economic recovery".
6. **Stock exchanges**: Here, too, one finds ongoing media coverage, the essential component of which are assumptions (or hypotheses) about the reasons for current price developments, for example, "Falling interest rates lead to rising stock prices".
7. **Management**: When assessing the performance of managers, one has to assume a (direct or indirect) cause-and-effect relationship between their actions and decisions, on the one hand, and the resulting effects on success, on the other hand.
8. **Marketing**: An example of (assumed) causal relationships in marketing decisions is the so-called realization of a sales promotion action (e.g., temporary price reduction). How could someone be responsible for the use of resources if he or she did not assume a causal link to a short-term increase in sales (causal chain: sales promotion → stimulation of customers to trial purchases and brand change → increased sales)?

Such considerations of causality have become quite natural to us. What are typical similarities of such (and, of course, other) causal relationships? Which characteristics entail causal relationships and then (logically) serve to decide in an empirical investigation whether a causal relationship exists or not? The first aspect relates to the **common variation of cause and effect**. Example 4 above shows that elevated blood pressure is associated with an increased risk of infarction, and in

Example 8 it is shown that increased sales promotion is associated with higher sales. In connection with the first feature is the possibility of **intervention** or **manipulation** of the (assumed) cause with the aim of achieving the desired (and assumed) effect. For instance, in Example 1, one might think about changing the behavior of parents through education or communication to attain better academic achievements of the children. In Example 4, the term "therapy" includes the attempt to eliminate the causes of a disease. As for Example 5, there are examples in the tax and subsidy policies of governments and the interest rate policy of central banks. However, there are causal relationships where such interventions are not possible. The third typical feature is the **temporal sequence** in the sense that the change of the (presumed) cause precedes the (presumed) effect. This may be a time interval in the range of seconds (e.g., in the case of a traffic accident caused by human error, see Example 2) or in the range of years (e.g., in the case of long-term damage to a bridge in Example 3). Fourthly, one assumes **the absence of alternative explanations** whose securing represents an essential and often complex problem in empirical research. Thus, in Example 1 poor academic performance could also be caused by teachers, in Example 3 the bridge could also have collapsed due to poor quality of the concrete and in Example 8 the sales figures could have increased because general demand has grown in the respective market. Only if one can exclude such (other) possible reasons for the observed effect, then can it be assumed that this effect is unmistakably caused by the assumed cause. Ultimately, there must be a **meaningful theoretical relationship** between cause and effect. Even if, in Example 6, one could observe a commonality of fluctuations of the outside temperature and the stock exchange market development—with a temperature increase regularly preceding a positive development of the stock prices, and no other possible causes for the price fluctuations being detected—still, hardly anyone would assume a causal relationship between temperature and the stock market. The following section intends to shed more light on these five aspects.

There is one important difference between the above eight examples. In some of the examples, the causal relationship relates to *specific cases,* while in others, more *general relationships are involved.* For instance, in the above examples—in law (2), there are typically case-related findings on guilt and responsibility, in medicine (4) diagnoses are made for individual patients, and individual evaluations are made of managers' performance (7). On the other hand, Examples 3, 6 and 8 refer to causal relationships, which have more general validity beyond individual cases. Nancy Cartwright (2014) distinguishes between **singular** and **general causal relationships**. In the sciences that focus on the development and testing of theories (see Sect. 2.1), interest in general causal relationships is greater. However, in some sciences (for example, in the science of history) the focus on important individual cases plays a major role (e.g., "What were the causes of World War I?"). In addition, the analysis of individual cases may also be helpful in other disciplines in the early stages of research (see Sect. 4.3.3). In the present chapter, however, general relationships are at the center of interest, since the test of causal hypotheses (typically through experiments, see Sect. 8.3) is oriented towards general causal relationships.

Now for the first feature of causal relationships, the **common variation of cause and effect**. Causal relationships are most likely to appear when the cause and effect *vary together*. If, for example, one observes several times that interest rates fall and then economic growth occurs, then this indicates a corresponding (causal) relationship. Remember, *this speaks in favor* of a causal relationship, but it is *not evidence* of a causal relationship. If interest rates and economic growth remain constant, then no evidence of a relationship exists and if the growth changes with interest rates remaining constant, then this speaks against a relationship. A change in the cause leads to a change or a difference in the effect (Psillos 2002, p. 6).

How can we imagine the relationship between cause and effect? In science and technology, one often encounters deterministic relationships, i.e., the effect always occurs (under all conditions such as location, situation, time, etc.) after the occurrence of the cause—often in a precisely determinable manner; for example, at reduced temperature, the resistance of an electric cable decreases. Such types of relationships hardly exist in the social sciences (including marketing research). Here, statements about probabilities or (with sufficiently large numbers of cases), statements about (relative) frequencies or correlations are more common. Nancy Cartwright (2014, p. 312) summarizes the basic idea: "When a cause is present there should be more of the effect than if it were absent. That is the root idea of the probabilistic theory of causation".

This way of establishing the relationship between cause and effect hardly differs from the analysis of relationships between variables discussed in the context of hypothesis testing in Chap. 7. Accordingly, to provide evidence for a causal relationship further requirements (see below) need to be met. Common variation of cause and effect is therefore a necessary, but by no means a sufficient, condition for a causal relationship. The well-known principle of **correlation ≠ causality** applies. With regard to causality, however, it is possible to ascertain that there is no causal relationship in the absence (or non-significance) of a correlation (or other measures or relationships).

The second aspect, the possibility of **intervention/manipulation**, has important practical and methodological consequences. On the one hand, it involves the use of knowledge of causal relationships for *design tasks*: in the examples given at the beginning of this section, Example 3 measures for the construction of a bridge, Example 4 for the determination of a therapy, Example 5 for an economic policy intervention and Example 8 for the realization of a promotional activity. Causal relationships are thus in a sense "recipes": If one understands a causal relationship, then one can shape causes in such a way that certain effects are achieved or prevented (Psillos 2002, p. 6). In *empirical investigations*, typically in experiments, the manipulation of independent variables and the observation of whether the dependent variables change in the expected manner are "classic" approaches (see Sect. 8.3). However, there are causal relationships in which this kind of observation and analysis is not possible. For example, while historians may ask for the causes of a particular event, they cannot test their assumptions through manipulation; the same is true for astronomers. In the social sciences, there are also some situations in which the manipulation of an independent variable is not possible (too much effort, high risk) or is ethically unacceptable (e.g., because of psychological or physical harm to study subjects). In such cases one often tries to come to comparable results by means of so-called quasi-experiments (see Sect. 8.3.3).

There is an interesting relationship of the previous paragraph to a fundamental aspect of various philosophy of science basic positions mentioned in Sect. 3.1, which deals with the position of **realism**, on the one hand, and **constructivism**, on the other. If one does not assume (in a constructivist view) that a reality exists that is independent of the viewer's perceptions and interpretations, then it makes little sense to carry out experiments. Under this assumption, the manipulation of real phenomena could have little impact on concepts and theories that exist only in the minds of scientists and have little to do with reality.

Theodore Arabatzis (2008, p. 164) explains the conflict between the constructivist view and the experimental approach:

"According to the early and most radical version of social constructivism, the constraints of nature on the products of scientific activity are minimal. Data are selected or even constructed in a process which reflects the social interactions within the relevant scientific community. Therefore, one should not appeal to the material world to explain the generation and acceptance of scientific knowledge."

The third characteristic of causality is the sequence of events in the form of **cause before effect**. Which one of the variables in a causal relationship is considered the "cause" and which one the "effect" has to be based on substantive considerations. Nevertheless, the answer is not always clear. For instance, a positive correlation between advertising expenditure and company profitability could either refer to the fact that advertising expenditure influences profitability or that profitability (by means of increased financial means) influences advertising expenditure. Here, the analysis of the *temporal sequence* can clarify matters. Basically, one assumes that the suspected cause occurs before the effect. If one observed in the example that first the advertising budgets increase and later profitability occurred, this speaks of a causal relationship "advertising expenses → profitability". Although Hitchcock (2008) refers to some special cases in physics in which the chronology and the direction of causality do not coincide, in the field of social science such an altered sequence is not quite conceivable. This also applies to cases in which certain expected events (e.g., expectation of a new iPhone, price developments) are anticipated and responded to, because in such cases the reactions are not due to these (often quite vague) future events, but due to the previously existing conjectures.

The central idea of the fourth feature, **absence of alternative explanations**, is quite simple and plausible. If one suspects a specific cause of an effect and is able to exclude all other possible causes as alternative explanations, then only the suspected cause remains to explain the effect. Alternative explanations can be both substantial and methodical. For example, reasons for a change in attitudes among consumers might be the impact of marketing communication, a change of values, or new experiences. However, the measured attitude change could also be due to a (systematic or random) measurement error. Researchers are usually not able to exclude all conceivable alternative explanations for a finding. Nevertheless, the research design should be designed in such a way that at least the most important alternative explanations (including the methodological ones) cannot play a role. In this context, keeping the influencing variables constant and using experimental and control groups plays an essential role in such study designs (see Sect. 8.3). By using experimental (with the presumed "cause") and control groups (no effect of the presumed "cause") and interpreting the results in the comparison of both groups, one achieves a situation where other predictors act in the same way in both groups. The difference between the group results can be attributed to the effect of the "cause". The prerequisite for this, however, is that there are no systematic differences between the two groups, which is generally achieved by randomizing the group assignment.

One type of causal relationship in the form of the so-called **INUS condition** explicitly takes into account the possibility that multiple causes and specific conditions for an effect may exist. This may be more in line with many marketing research questions than a simple relationship of just one possible cause and effect. "INUS" is an abbreviation for **I**nsufficient–**N**ecessary–**U**nnecessary–**S**ufficient (see, for example, Bagozzi 1980, pp. 16ff., Psillos 2002, pp. 87ff.). What is meant by this (initially somewhat cryptic) name? "A cause may be an insufficient but necessary part of a condition that is itself unnecessary but sufficient for the result" (Bagozzi 1980, p. 17). Since the central idea might still not be easy to understand, here is an example of the following causal relationship: "Advertising messages change attitudes":

- *Not necessary for the result*: Changes in attitudes can be due to other causes (e.g., consumer experiences). Hence, advertising is not necessary for changes in attitudes.
- *Insufficient part of the conditions*: Advertising messages alone do not change any attitudes (are therefore not sufficient), but it is only under the conditions that consumers are exposed to the message, that they show sufficiently high involvement, etc.
- *Sufficient for the result*: If the conditions (see above) apply, then the attitude change arises as an effect of advertising messages; advertising would be sufficient *under these conditions*.
- *Necessary part of the conditions*: If the advertising message did not exist, then under the given conditions, attitudes would not change. Hence, advertising would therefore be necessary in this context to change attitudes.

Figure 8.1 graphically illustrates the example of an INUS condition as outlined above.

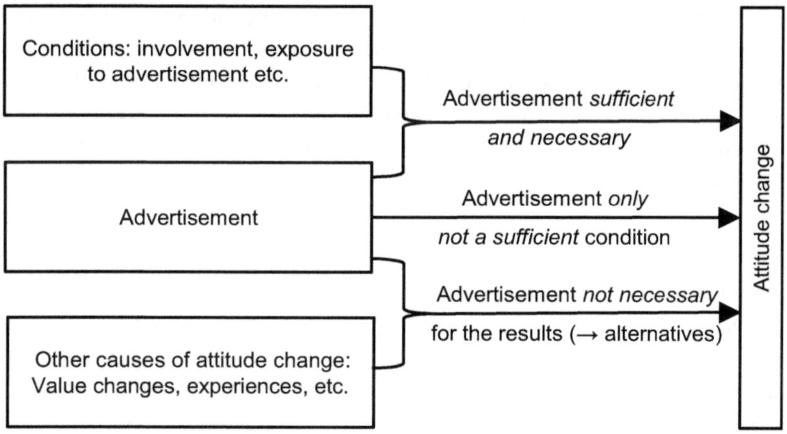

Fig. 8.1 Example of INUS conditions

Another example of Psillos (2004, p. 277) may further illustrate the somewhat complicated INUS condition:

"To say that short circuits cause house fires is to say that the short circuit is an INUS condition for house fires. It is an insufficient part because it cannot cause the fire on its own (other conditions such as oxygen, inflammable material, etc. should be present). It is, nonetheless, a nonredundant part because, without it, the rest of the conditions are not sufficient for the fire. It is just a part, and not the whole, of a sufficient condition (which includes oxygen, the presence of inflammable material, etc.), but this whole sufficient condition is not necessary, since some other cluster of conditions, for example, an arsonist with gasoline, can produce the fire."

Let us now go back to the characteristics of causal relationships. Here is the fifth feature, where the relationship **should have a theoretical foundation.** The word "causal" already suggests that it is not about random relationships, but systematic and well-founded relationships between variables. In the social sciences, therefore, it is common to develop a chain of causation that explains and justifies the relationship between cause and effect (Cartwright 2014). For example, such a causal chain in the above described relationship between advertising and attitude change might look like this: advertising appears on TV → consumer watches and receives the message → message evokes cognitive and/or emotional responses → change of previous beliefs and evaluations → attitude change. An empirical way of analyzing such causal chains are so-called mediators, which will be discussed in Sect. 8.2.

However, with regard to the demand of a theoretical justification for a causal relationship, it should be kept in mind that this could intensify the problem of the theory-ladenness (see Sect. 3.2 and Arabatzis 2008). Corresponding empirical studies (experiments) are typically based on previously theoretically based hypotheses and are designed accordingly. This relates to the perception and interpretation of results by the researchers, who in most cases are also "followers" of the respective theory and often try to confirm it. Peter (1991) also points out that in research practice (occasionally? often?) a research design undergoes several pretests and changes until the desired result appears, which, of course, can be problematic from an ethical research perspective (see Sect. 10.2.2).

David de Vaus (2001, p. 36) explains why a theoretical justification for the assumption of a causal relationship is essential:

"The causal assertion must make sense. We should be able to tell a story of how X affects Y if we wish to infer a causal relationship between X and Y. Even if we cannot empirically demonstrate how X affects Y we need to provide a plausible account of the connection (plausible in terms of other research, current theory etc.)."

Of the five characteristics of a causal relationship, only one—the common variation of cause and effect—directly affects the methods of statistical analysis, because it is a question of (significant) differences and changes. The last feature, the requirement of a theoretical foundation, is outside the methodological area. The three other features (manipulation, time sequence of cause before effect, and absence of alternative explanations) primarily concern the study design. "The ability to make a causal inference between two variables is a function of one's research design, not the statistical technique used to analyze the data that are yielded by that research design" (Jaccard and Becker 2002, p. 248). Empirical methods for verifying causal relationships are typically experiments because there is close correspondence between the five outlined criteria for a causal relationship and the central elements of experimental design in experiments (see Sect. 8.3). Therefore experiments can test assumptions about causal relationships, i.e., causal hypotheses.

8.2 Types of Causal Relationships

The examination of *causal hypotheses* places particularly high demands on the methodological procedure. They lead to substantial statements in science and practice. If a researcher has determined that a particular combination of mental traits is the cause of a particular work behavior, then he or she has come a good deal closer to the goal (at least from the perspective of scientific realism) of understanding and explaining reality. When a product manager finds that certain product quality problems are the cause of decreasing market shares of a product, then he or she has found a critical starting point to solve the problem of decreasing market share.

In Fig. 8.2 there is an overview of different types of relationships between variables that either mimic causal relationships or misinterpret causal relationships.

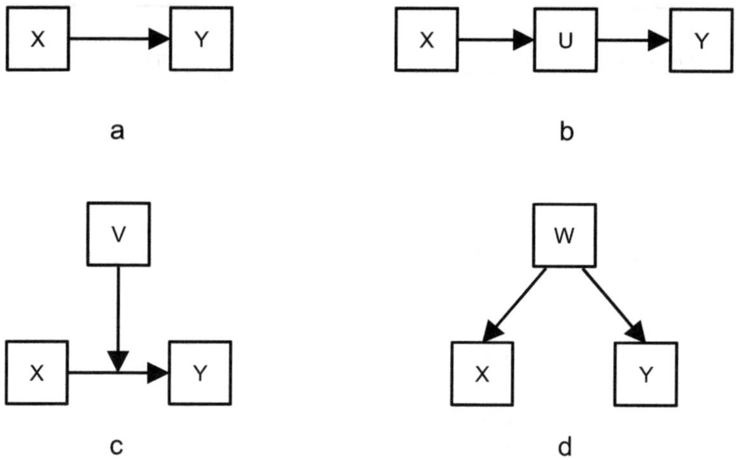

Fig. 8.2 Types of (causal) relationships between variables. (**a**) Direct causal relationship. (**b**) Indirect causal relationship. (**c**) Moderated causal relationship. (**d**) Spurious relationship (see e.g., Jaccard and Jacoby 2010)

Part a shows a simple, **direct causal relationship**, for example, the contact with advertising (cause) on the attitude to a product (effect). Part b shows an indirect causal relationship with a mediator variable (for explanation, see below). Part c shows a moderated causal relationship in which the effect of X on Y is influenced by a third variable, V (see below for explanation). Finally, part d shows a relationship that does not represent a causal relationship between X and Y because a common variation of X and Y is caused by a third variable, W. For example, the common variation of income and use of print media can be under the influence of a third variable, education. There is a danger here that the relationship between X and Y could be misinterpreted as a causal relationship.

In the moderated causal relationship, the **moderator**, a second independent variable, moderates the effect of an independent variable on a dependent variable. The influence of the independent on the dependent variable becomes stronger or weaker. The moderator can also reverse the direction of the influence: "a moderator is a qualitative (e.g., sex, race, class) or quantitative (e.g., level of reward) variable that affects the direction and/or strength of the relation between an independent or predictor variable and a dependent or criterion variable" (Baron and Kenny 1986, p. 1174). As an example, one might think of the above relationship between exposure to advertising (X) and attitude to a brand (Y), which is moderated by the involvement with the product category: the more a consumer is involved with a product category, the stronger the effect of the exposure to advertising will be on attitudes towards a brand in that product category.

Mediators differ from moderators. Mediators designate indirect relationships between variables. Figure 8.3 shows a well-known example from advertising research (MacKenzie et al. 1986). The idea is that advertising influences attitudes towards the advertised brand. This acts, on the one hand, as a direct effect, but can

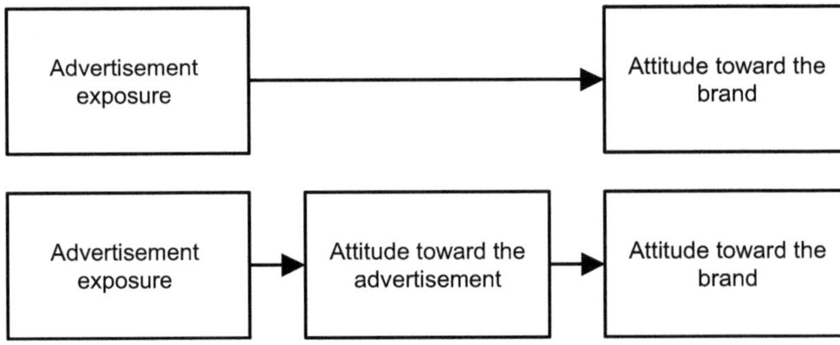

Fig. 8.3 Example of a mediator and an indirect causal relationship

also be explained by attitude to the advertisement as an indirect effect: advertising leads to the changes in attitude to the advertisement, which in turn changes the attitude to the brand. Both relationships can theoretically be justified. A direct relationship in one view (or theory) can therefore be an indirect relationship in another view (or theory).

8.3 Experimental Studies

8.3.1 Nature and Design of Experiments

Due to the five requirements for establishing causal relationships explained in Sect. 8.1, a particular study design, known as **experiment**, is commonly used. In essence, an experiment is an approach in which one or more independent variables are manipulated in such a way that the corresponding effects on a dependent variable can be observed. It is therefore a question of determining whether a certain (independent) variable is actually the reason (the cause) for a change of another (dependent) variable (effect).

Typical of experiments is the isolated consideration of the variables of interest. One does not look at a variety of factors influencing, for instance, a decision and their interactions, instead the experiment focuses only on the influence of a particular element in advertising (e.g., color or music) on the attitudes of consumers. For this reason, experimental investigations often reveal a certain *artificiality of the research design*, which is based on the exclusion of other influencing factors (→ absence of alternative explanations). Against this background, it is also easy to understand that today, one can find the results of more than one empirical study in many publications in which experiments are used. In each study, individual aspects are considered in isolation and the resulting summaries constitute a more comprehensive investigation of a topic.

Alan Chalmers (2013, p. 26) illustrates the intention of an isolated observation in the context of experiments by using the following example:

"Many kinds of processes are at work in the world around us, and they are all superimposed on, and interact with each other in complicated ways. A falling leaf is subject to gravity, air resistance and the force of winds and will also rot to some small degree as it falls. It is not possible to arrive at an understanding of these various processes by careful observation of events as they typically and naturally occur. Observation of falling leaves will not yield Galileo's law of fall. The lesson to be learned here is rather straightforward. To acquire facts relevant for the identification and specification of the various processes at work in nature it is, in general, necessary to practically intervene to try to isolate the process under investigation and eliminate the effects of others. In short, it is necessary to do experiments."

The major conclusions in experimental investigations can be explained by the example of a "classical" experimental design according to de Vaus (2001, pp. 48–49). The following features characterize this design:

- A pre-measure (→ sequence of cause and effect)
- Two groups: experimental group and control group (→ absence of alternative explanations)
- Random assignment of the subjects to the two groups (→ absence of alternative explanations)
- An intervention (manipulation)
- A final measurement (→ order of cause and effect)

Table 8.1 illustrates such a design. It shows the measurement times, the assignment of subjects to groups and the intervention. In both groups, attitude to a brand is pre-measured. Then, only the subjects in the experimental group are confronted with advertising for the brand. This is the **intervention** or **manipulation** of the independent variable. In the example shown, the manipulation is carried out very simply by confronting the experimental group with advertising, but not the control group. Manipulations can be diverse and can even affect mental states (such as motivations

Table 8.1 Example of a classical experimental design (according to De Vaus 2001, p. 49)

Random assignment to experimental groups	Pre measure t_1	Intervention (manipulation of independent variable) t_2	Post measure t_3
Treatment group	Attitude toward the brand t_1	*Exposure* to advertisement	Attitude toward the brand t_3
Control group	Attitude toward the brand t_1	*No exposure* to advertisement	Attitude toward the brand t_3

or emotional states). For this purpose, the different groups of subjects are influenced (or manipulated) in such a way that the corresponding mental states occur among the members of the various groups. For example, one could achieve different levels of motivation through different incentives. This process of **operationalization** (see Sect. 6.1) aims to achieve different values of independent variables. Therefore Aronson et al. (1998, p. 111) speak of "constructing the independent variable". **Manipulation checks** usually control whether these manipulations have succeeded (e.g., whether the motivation or emotional state differs between the experimental groups). After the intervention or manipulation, the attitude to the brand is measured once more. This can occur verbally (through the use of a questionnaire) or through observations. As in the case of manipulation, one needs to consider the aspects and quality criteria of operationalization. If a significant change of attitude is measured in the experimental group only, then one would consider it as being caused by the contact with the advertisement. Are the conditions outlined above for a causal relationship given in this example?

The example fulfills the conditions for a causal relationship, if the corresponding empirical results show the expected values. This can be shown as follows:

- *Common variation of cause* (in the example "exposure to advertisement") *and effect* (in the example "attitude to the brand" at time t_3): This condition is clearly fulfilled, since the intervention in the form of the contact with the advertisement takes place only in the experimental group. The contact with the advertisement thus varies between the groups and its measurement shows whether the dependent variable alters between both experimental groups accordingly.
- An intervention/manipulation at time t_2 is part of the experimental design.
- *Change of the cause* (in the example: exposure to the advertisement) *before change of the effect* (in the example: attitude change): This requirement is also fulfilled by the experimental design, which determines the timing of intervention and post-measure.
- *Absence of alternative explanations*: In field studies, the exclusion of all conceivable alternative explanations can hardly ever be achieved. This is certainly a weak point of experiments. Therefore, one focuses on particularly important or frequently occurring aspects of an investigation. Of central importance is the use of (comparable!) **experimental and control groups**. Ideally, these groups do not differ except for the intervention (e.g., they do not differ in terms of socio-demographic or psychological characteristics, past experience and attitudes). Therefore, different results of the final measure can only be attributed to the "cause" in the form of the intervention. In most cases, the assignment of subjects to experimental and control groups is random (**randomization**), which makes greater differences between the two groups less likely. In the example shown, the random assignment of the subjects to the experimental and control groups has (largely) excluded the fact that these groups differ systematically from one another, which could be an alternative explanation for differences in the final measure. For this reason, researchers like to work with students as subjects in experiments, because this group is largely homogeneous in terms of many

demographics (e.g., age, education, income) as well as psychographic characteristics (e.g., openness to innovation), which further reduces the risk of systematic differences. As mentioned in Sect. 6.4, however, experiments with students may be problematic if generalizability of the results is desired, but the results are systematically different from the population, for example, if students are generally more positive about advertising. Then, in the example mentioned above, students may experience an effect that may not be present in other people (non-students) or that is not so strong. Due to randomization, a pre-measurement is no longer necessary, because one can assume that the attitude to the brand at time t_1 is randomly distributed over both groups and thus on average should be approximately the same in both groups. When interpreting the results of the study, one focuses on statistically significant differences between the groups and neglects random (small) group differences with regard to the hypothesis of the investigation. Randomization as *random assignment* to experimental or control groups should be clearly differentiated from the *random selection of subjects (random sample)*, which in experiments serve in particular to achieve external validity (see Sect. 8.3.2).

The above-mentioned alternative explanations, which are based on the methodological procedure in an experiment, are discussed in the following Sect. 8.3.2 under the heading "internal validity". The rather complex design of experimental studies typically aims to exclude several alternative explanations (see, e.g., Shadish et al. 2002; Koschate-Fischer and Schandelmeier 2014).

- *Theoretical justification of the relationship*: The methodology cannot answer the question as to whether there is an adequate theoretical justification for an examined relationship, but a substantive consideration can. The development of an experimental design forces researchers to make deliberate considerations regarding the mode of action of independent and dependent variables (i.e., corresponding theoretical considerations). In the example used here (advertising → attitude change), the theoretical justification is established and easy to understand.

Experiments have long been widely used and are accepted methods in medicine or psychology. Accordingly, psychology-related areas of marketing research use them quite frequently (in particular, consumer research). The applications of experimental designs are typically more complex than the example given. They often examine two or three independent variables at the same time, as well as their interactions, and make manifold efforts in order to meet the requirements for the examination of causality. Please refer to the extant literature (e.g., Koschate-Fischer and Schandelmeier 2014; Shadish et al. 2002; Geuens and Pelsmacker 2017).

8.3.2 Internal and External Validity of Experiments

Chapter 6 explained the importance of the reliability and validity of a study with regard to the meaning of study results. As already mentioned, the problem is that

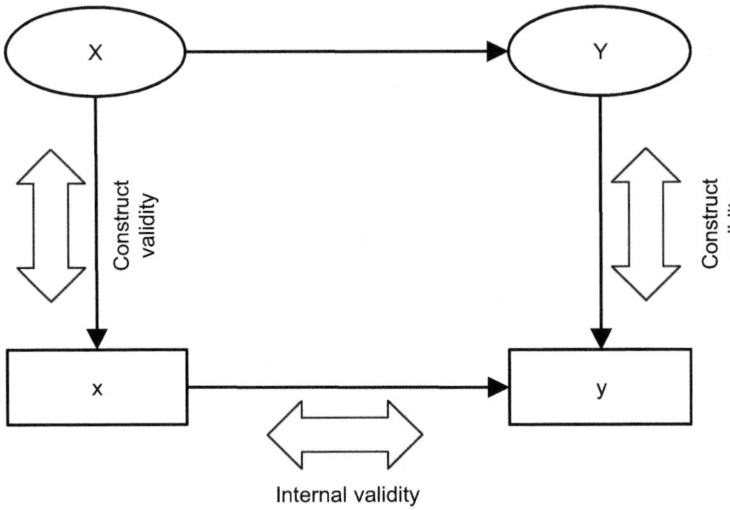

Fig. 8.4 Internal validity and construct validity in experiments

results that confirm or do not confirm a hypothesis are limited in their validity in terms of the theory tested, if these results are influenced by method errors. Concerning experiments, general considerations on the validity of studies (see Sect. 6.3) add two specific aspects: internal and external validity. The aspect of internal validity has already been implicitly addressed. **Internal validity** refers to the *elimination of alternative explanations* for the observed relationships due to the measurement process. Internal validity is thus "the validity of inferences about whether the relationship between two variables is causal" (Shadish et al. 2002, p. 508). The main question here is whether the change in a dependent variable can actually be attributed to the presumed cause, i.e., the change in an independent variable, or whether inadequacies of the methods and the measurements are responsible for the results. Figure 8.4 shows this aspect and the relation of the measured variables to the theoretically interesting concepts/constructs (→ construct validity, see Sect. 6.3.3). The lower-case letters (x, y) stand for the variables used in the study, which should be an operationalization of the corresponding concepts/constructs (upper-case letters X, Y). Construct validity is primarily related to validity in the measurement of concepts (has the concept been measured correctly?), the internal validity is concerned with the question of whether the relationship between concepts is validly represented (does the measured relationship actually exist?).

The internal validity of an experiment is mainly jeopardized by the problems mentioned below (Shadish et al. 2002, pp. 54ff.). They provide alternative explanations for the results of experiments, which are methodologically justified and that should be avoided by the design of the experimental design.

- *Selection/assignment.* The assignment to experimental and control groups might not ensure that neither group shows any systematic differences. Thus, if a difference exists between the groups, one cannot infer the effect of the independent variables.

- *History*. Each event between pre- and post-measure may have an unwanted influence on the subjects, such as external influences that affect only a part of the subjects.
- *Maturing*. Subjects can change between two measures due to experience, fatigue etc. Therefore, it could be that subjects respond differently to stimuli over time and thus their actual effect is mitigated or nullified.
- *Change in measurement instruments*. During a study, the characteristics of the measurement instruments, including the measuring persons, may change. For example, the measurements may be made more accurate by increasing the experience of the measuring persons, or less accurate by increasing boredom during the course of the experiment.
- *Regression to the mean*. This statistical artifact can be superimposed on effects, for example, by selecting subjects with particularly extreme values, who then show (as a statistical necessity), on subsequent measures, quite "moderate" values.
- *Drop out*. Subjects may drop out during the study due to the study requirements. The affected groups are then smaller in a second measurement, which in turn can influence the result in case of a non-random drop-out.

In addition, the question arises to what extent the results of a study can be generalized. What explanatory power, for example, does a study that was carried out on German product managers have for product managers in general? What do the results of a consumer behavior experiment with 100 American students say about consumers in general? Such questions apply to the external validity of experiments. **External validity** refers to the generalizability (see also Chap. 6) of results about different persons, situations, contexts etc. External validity is therefore: "the validity of inferences about whether the causal relationship holds over variations in persons, settings, treatment variables, and measurement variables" (Shadish et al. 2002, p. 507).

Campbell and Stanley (1963, p. 5) formulate the central points of internal and external validity as follows:

"Fundamental (. . .) is a distinction between internal validity and external validity. Internal validity is the basic minimum without which any experiment is uninterpretable: Did in fact the experimental treatments make a difference in this specific experimental instance? External validity asks the question of generalizability: To what populations, settings, treatment variables, and measurement variables can this effect be generalized? Both types of criteria are obviously important, even though they are frequently at odds in that features increasing one may jeopardize the other. While internal validity is the sine qua

(continued)

non, and while the question of external validity, like the question of inductive inference, is never completely answerable, the selection of designs strong in both types of validity is obviously our ideal."

The four main considerations of external validity are as follows:

- Can the results from the typically small number of subjects (for example, persons, companies) be transferred to corresponding populations? The answers to such questions usually lie in the tools of random sampling theory and inferential statistics (see Sect. 6.4).
- Is the generalization of the results possible with regard to the object of investigation (e.g., attitude to a product → attitude to a retailer)?
- Are the results transferrable to other contexts (for example, other cultural environments, other times)?
- Does one get the same results when using other methods of examination (such as other measurements) or do the results depend on the method?

The sources of danger for the external validity of experiments are (Shadish et al. 2002):

- *Biased selection.* Selecting participants in a way that they are not representative of the population under investigation weakens the generalizability of the results.
- *Reactivity of the experiment.* The manipulations in a controlled laboratory environment may not apply to a less controllable real environment.

With regard to practical issues, external validity is indispensable, because it is about making inferences from the results of a study on the events in broader contexts (e.g., markets) for which decisions are to be made (Calder et al. 1982). This also shows that the use of experiments is by no means limited to the examination of causal relationships in theories. Particularly in practice, questions often arise such as, "What would happen if ....?". The representative selection of test subjects (analogous to the typical procedure for representative surveys) and a realistic ("natural") examination situation obviously have special significance for the external validity. However, as discussed above, these two issues often present challenges to internal validity, where homogeneity of subjects and artificial testing situations are favored to minimize the influence of confounding factors. In the literature, there are extensive discussions on how to try to increase the realism of experiments without reducing the credibility of the results, i.e., to ensure external and internal validity at the same time (Geuens and Pelsmacker 2017; Morales et al. 2017). These include, above all, the design of realistic experimental stimuli, the use of behavioral variables as dependent variables, and the composition of the sample. Because there is a *trade-off between the internal and external validity of experiments*, achieving both goals at the same time is a challenging task and almost impossible to achieve.

8.3.3 Quasi-experiments

Typical for the above-identified experimental designs are the controlled (or manipulated) use of the independent variable and the random assignment of subjects to experimental and control groups. The aim is to eliminate systematic differences between these groups that might bias the effect of the independent variables. There are situations in which these conditions do not occur. Two examples may illustrate this problem:

- To investigate whether the children of smokers are more likely to become smokers than other people: it is obvious that a random assignment to the two groups to be compared ("parents are smokers" and "parents are non-smokers") is not only practically impossible, but also ethically highly questionable.
- To investigate whether home ownership affects budget allocation and consumer behavior over the long term (10 years or more): one will barely have 10 years to observe the consumer choice behavior of homebuyers in contrast to tenants. It would be more viable to find out from current homeowners and tenants what behavioral differences arise. That would certainly not be a random assignment, but would solve the problem of the duration of the study.

Campbell and Stanley (1963, p. 34) speak of **quasi-experiments** in situations in which essential principles of experimental investigations are applied without being able to meet all relevant requirements. There are a number of reasons for the necessity and application of quasi-experiments:

- A *randomized assignment* of subjects to the experimental groups is often *not possible*, for example, if one wants to check the effects of different viral infections.
- *Ethical reasons* often also speak against experimental manipulations, even if it were possible, such as in reviewing the effects of illegal drugs.
- The *duration of the experiment* can be too long to apply a classical experimental design, for example, in examining the long-term impact of the media on a society's values.

Quasi-experiments thus are characterized by the fact that a randomized assignment of subjects to the experimental groups is not possible; that an independent variable cannot be manipulated and that there are no interventions that influence the dependent variable of the study.

Campbell and Stanley (1963, p. 34) on quasi-experiments:

"There are many social settings in which the research person can introduce something like experimental design into his scheduling of data collection

(continued)

procedures (e.g., the when and to whom of measurement), even though he lacks the full control over the scheduling of experimental stimuli (the when and to whom of exposure and the ability to randomize exposures) which makes a true experiment possible."

Kerlinger and Lee (2000, p. 536) identify the reasons for carrying out quasi-experiments:

"The true experiment requires the manipulation of at least one independent variable, the random assignment of participants to groups, and the random assignment of treatments to groups. When one or more of these prerequisites is missing for one reason or another, we have a compromise design. Compromise designs are popularly known as quasi-experimental designs."

In quasi-experiments—by the necessary absence of the random assignment of study subjects to experimental and control groups—a *confounding and distorting effect cannot be excluded*, so other ways are necessary to assure the absence of alternative explanations. Shadish et al. (2002, p. 105) emphasize the "identification and study of plausible threats to internal validity" by critically examining potential alternative influencing factors, which are typically considered as additional *control variables* in data analysis. If, for example, one wants to check whether the (non-) smoking behavior of the parents has an influence on whether the children become smokers, then it makes sense to also include control variables that describe the social environment, or the children's personality, and provide alternative explanations. On the other hand, quasi-experiments often have advantages in terms of external validity, because the data were collected in "natural" situations.

8.4 Complex Causality

Causal hypotheses, as well as the analytical procedures for investigating causality, usually assume causal relationships that assume the necessary and sufficient conditions for an effect (for example, "the more investment, the more revenue"). **Complex causality** means distinguishing between *different forms of causality* by distinguishing between combinations of necessary and sufficient conditions. Schneider and Eggert (2014) illustrate four forms of causality, exemplifying the relationship between the two concepts of commitment and trust in a business relationship. This research assumes that trust leads to commitment in a business relationship, that is, trust is a cause, and commitment is the effect:

- One variable is a necessary but not sufficient condition for the occurrence of another variable. That is, commitment occurs when trust occurs, but does not need to, so that trust can occur without there being any commitment.

- A variable is a sufficient but not a necessary condition for a second variable. That is, commitment occurs when trust occurs, but commitment can also occur without trust.
- A variable can be part of a combination of sufficient conditions without itself being sufficient or necessary. Trust might explain commitment sufficiently well, but only in combination with other factors, such as the benefit of a relationship. Trust would then be a so-called INUS condition (see Sect. 8.1).
- One variable is a sufficient and necessary condition for the occurrence of a second variable. That is, trust always leads to commitment and commitment without trust does not occur.

The typical technique used to analyze complex causalities is **Qualitative Comparative Analysis (QCA)**. QCA is a method of causal analysis of configurational data in the social sciences. Configuration data means that all variables, no matter what measurement levels, are converted to qualitative data, for example, different levels of trust, which are typically measured as an interval-scaled variable, convert to "trust exists/trust does not exist". Furthermore, there is a difference between an "outcome", which in principle is the effect (here: commitment), as well as the "conditions", these are the causes and possible moderators (here: trust, benefit of a relationship, etc.). For each observation (e.g., for each business partner), a value between 0 and 1 is entered into a truth table for the conditions and the outcome, which indicates to what extent the observation tends towards one or the other characteristic of the configurational variables (e.g., the probability of the occurrence of trust or commitment). Subsequently, algorithms are applied, with the search objective to identify minimally necessary and sufficient conditions for the presence of the outcome: if, for example, in all observations in which commitment (the outcome) is found, there is always trust, then trust is a necessary condition for commitment. For the details of this analysis, please refer to the relevant literature (e.g., Ragin 2008; Schulze-Bentrop 2013). The result of the analysis indicates those conditions that are necessary and those that sufficiently explain the outcome. This can be a single condition, but it can also be combinations of conditions.

The advantage of QCA over other, non-experimental methods of causal analysis is the identification of the causes of an effect. However, if one wants to examine how much one particular variable (cause) contributes to the explanation of another variable (effect), then conventional regression-based analysis techniques are more appropriate.

References

Arabatzis, T. (2008). Experiment. In S. Psillos & M. Curd (Eds.), *The Routledge companion to philosophy of science* (pp. 159–170). London: Routledge.

Aronson, E., Wilson, T., & Brewer, M. (1998). Experimentation in social psychology. In D. Gilbert, S. Fiske, & G. Linzay (Eds.), *The handbook of social psychology* (4th ed., pp. 99–142). New York: McGraw-Hill.

Bagozzi, R. P. (1980). *Causal models in marketing*. New York: Wiley.

Baron, R. M., & Kenny, D. A. (1986). The moderator-mediator variable distinction in social psychological research: Conceptual, strategic, and statistical considerations. *Journal of Personality and Social Psychology, 51*(6), 1173–1182.

Calder, B., Phillips, L., & Tybout, A. (1982). The concept of external validity. *Journal of Consumer Research, 9*(3), 240–244.

Campbell, D. T., & Stanley, T. D. (1963). *Experimental and quasi-experimental designs for research.* Chicago: Rand McNally.

Cartwright, N. (2014). Causal inference. In N. Cartwright & E. Montuschi (Eds.), *Philosophy of social science* (pp. 208–326). Oxford: Oxford University Press.

Chalmers, A. (2013). *What is this thing called science?* (4th ed.). Indianapolis, IN: Hackett.

De Vaus, D. (2001). *Research design in social research.* London: Sage.

Geuens, M., & Pelsmacker, P. D. (2017). Planning and conducting experimental advertising research and questionnaire design. *Journal of Advertising, 46*(1), 83–100.

Godfrey-Smith, P. (2003). *Theory and reality—An introduction to the philosophy of science.* Chicago: University of Chicago Press.

Hitchcock, C. (2008). Probabilistic causation. In E. N. Zalta (Ed.), *The Stanford encyclopedia of philosophy.* http://plato.stanford.edu

Humphreys, P. (2000). Causation. In W. Newton-Smith (Ed.), *A companion to the philosophy of science* (pp. 31–40). Malden, MA: Wiley.

Jaccard, J., & Becker, M. (2002). *Statistics for the behavioral sciences* (4th ed.). Belmont, CA: Wadsworth.

Jaccard, J., & Jacoby, J. (2010). *Theory construction and model-building skills—A practical guide for social scientists.* London: Guilford.

Kerlinger, F., & Lee, H. (2000). *Foundations of behavioral research* (4th ed.). Melbourne: Wadsworth.

Koschate-Fischer, N., & Schandelmeier, S. (2014). A guideline for designing experimental studies in marketing research and a critical discussion of selected problem areas. *Journal of Business Economics, 84*(6), 793–826.

MacKenzie, S. B., Lutz, R. J., & Belch, G. E. (1986). The role of attitude toward the ad as a mediator of advertising effectiveness: A test of competing explanations. *Journal of Marketing Research, 23*(2), 130–143.

Morales, A. C., Amir, O., & Lee, L. (2017). Keeping it real in experimental research—Understanding when, where, and how to enhance realism and measure consumer behavior. *Journal of Consumer Research, 44*(2), 465–476.

Mumford, S., & Anjum, R. (2013). *Causation—A very short introduction.* Oxford: Oxford University Press.

Peter, J. (1991). Philosophical tensions in consumer inquiry. In T. Robertson & H. Kassarjian (Eds.), *Handbook of consumer behavior* (pp. 533–547). Englewood Cliffs, NJ: Prentice-Hall.

Psillos, S. (2002). *Causation & explanation.* Durham: Acumen.

Psillos, S. (2004). Causality. In M. Horowitz (Ed.), *New dictionary of the history of ideas* (pp. 272–280). New York: Charles Scribners.

Ragin, C. (2008). *Redesigning social inquiry: Fuzzy sets and beyond.* Chicago: University of Chicago Press.

Schneider, M. R., & Eggert, A. (2014). Embracing complex causality with the QCA method: An invitation. *Journal of Business Market Management, 7*(1), 312–328.

Schulze-Bentrop, C. (2013). *Qualitative comparative analysis (QCA) and configurational thinking in management studies.* Frankfurt am Main: Peter Lang.

Shadish, W. R., Cook, T. D., & Campbell, D. T. (2002). *Experimental and quasi-experimental designs for generalized causal inference.* Boston: Houghton Mifflin.

Further Reading

Geuens, M., & Pelsmacker, P. D. (2017). Planning and conducting experimental advertising research and questionnaire design. *Journal of Advertising, 46*(1), 83–100.

Koschate-Fischer, N., & Schandelmeier, S. (2014). A guideline for designing experimental studies in marketing research and a critical discussion of selected problem areas. *Journal of Business Economics, 84*(6), 793–826.

Psillos, S. (2002). *Causation & explanation*. Durham: Acumen.

Shadish, W. R., Cook, T. D., & Campbell, D. T. (2002). *Experimental and quasi-experimental designs for generalized causal inference*. Boston: Houghton Mifflin.

Generalizability of Research Results 9

9.1 Empirical Research and Generalizability

Section 2.5 outlines ways of gaining scientific knowledge. There are different approaches used to develop new theories, among them the attempt to generalize from observations (→ induction, see Sect. 2.5). This is followed by tests and, if necessary, negation or modification of the theory.

No matter how many observations are the same, one cannot draw *definite* conclusions as to corresponding lawlike generalizations. At some point, an unknown and deviant case can occur. On the other hand, one of the central aspects of scientific realism, as noted in Sect. 3.1, is reminiscent:

If a theory and the statements contained in it prove themselves long-term and often in appropriate tests and in practical applications, then there is obviously much to suggest that these statements are relatively likely to be approximately true, although, of course, one cannot achieve any certainty.

This chapter deals with approaches in which different findings about the same research question are summarized (→ meta-analyses) or new studies are carried out to check previous results (→ replications). Such approaches focus on the **generalizability** of research results.

The generalizability of test results relates to the question of how well one can extrapolate from a particular result to other *subjects* (e.g., sample → population), *research objects* (e.g., success of companies during the introduction of a product → success of companies in general), *contexts* (e.g., USA → Europe, present → future) and when using other *methods* (e.g., laboratory experiment → field study).

M. Eisend, A. Kuss, *Research Methodology in Marketing*,
https://doi.org/10.1007/978-3-030-10794-9_9

Kerlinger and Lee (2000, p. 474) characterize the importance of generalizability as follows:

"Can we generalize the results of a study to other participants, other groups, and other conditions? Perhaps the question is better put: How much can we generalize the results of the study? This is probably the most complex and difficult question that can be asked of research data because it touches not only on technical matters (like sampling and research design), but also on larger problems of basic and applied research."

Of course, the ability to generalize results is critical to many application-oriented investigations. The results of a customer survey are usually only relevant if one can apply the results to the behavior of customers in general. Typically, the focus in application-oriented research is primarily about generalizing from a sample to a population of interest.

We previously presented one generalizability approach regarding the quality criteria of measurement in Sect. 6.3. Here, one aims at the generalizability of a measurement by reducing measurement errors. This is typically a question of reliability, whereby classic reliability tests take into account only one source of error (for example, time of study). A consideration of different measurement errors that can occur simultaneously, as well as their interaction, is performed in the context of **generalizability theory** (Cronbach et al. 1972, see also Sect. 6.3.4). This chapter is not about the generalizability of measurements, but, in line with the previous chapters about theory and hypotheses testing, about the **generalizability of research results**, which usually refers to the relationships of variables according to the proposed hypotheses. Of course, the validity, reliability, and generalizability of measurement instruments is an essential prerequisite for the generalizability of test results, since the results can only be meaningfully interpreted if the measurement error is as small and controllable as possible. In scientific research, replication studies and meta-analyses are commonly carried out in order to arrive at generalizable statements regarding study results.

9.2 Replication Studies

Replication studies are repetitions of empirical studies aimed at demonstrating the reproducibility of the results. Replication studies do not differ from the original studies in terms of the study object or research question, but often in some aspects of the procedure. Through (successful) replication studies, one can achieve a certain independence for the study's results from sampling errors, the specifics of the research methods, and, at a minimum, from the time the original study was conducted. Here, we again consider the requirements for inductive reasoning mentioned in Sect. 2.5—that observations under different conditions should lead to the same result.

In general, the possibility of replication or the reproducibility of results of empirical studies is an *essential criterion for the scientific value* of studies. Therefore, it is common in the natural sciences that results must be replicable. This achieves the independence of the results of a particular study context and study method, as well as some protection against results that may be biased by the research process and the researchers (see Sect. 10.2). In the context of the discussion of paradigms and relativism (Chap. 3), it was suggested that empirical results could also be systematically influenced by the view of the researchers (theory-ladenness) and by the applied methods. By replicating the use of different methods performed by different researchers, independence from such influences is more likely to be ensured. McCullough and Vinod (2003, p. 888) describe the replicability of studies as a basic requirement of science: "Replication is the cornerstone of science. Research that cannot be replicated is not science, and cannot be trusted either as part of the profession's accumulated body of knowledge or as a basis for policy."

Hunter (2001) distinguishes the following *types of replications*:

- **Statistical replications** refer to exact repetitions of previous studies with the aim to increase the accuracy of statistical results by reducing the sampling error.
- **Scientific replications** refer to studies that use equivalent but not identical methods when repeating previous studies.
- **Conceptual replications** are replication studies with deliberate changes made to the original study. The change occurs, for instance, by including additional variables for the purpose of examining further potential influencing factors or by so-called moderator variables, which either limit or generalize the scope of the previous findings.

Kerlinger and Lee (2000, p. 365), on the nature and significance of replication studies, state:

"Whenever possible, replicate research studies...The word replication is used rather than repetition because in a replication, although the original relation is studied again, it might be studied with different kinds of participants, under somewhat different conditions, and even with fewer, more, or even different variables."

Although replicability of studies is obviously an important prerequisite for the scientific acceptance of research results, replication attempts often fail. For example, in a large-scale replication project, hundreds of studies published in leading psychology journals were replicated (Open Science Collaboration 2015). Only 36% of the studies showed an effect consistent with the effect of the original study as the effects in the replication studies tended to be weaker than the effects in the original studies. This indicates the existence of a publication bias (see Sect. 9.3).

Also, in marketing research, *the success rates of replication studies* are similarly low (e.g., Hubbard and Vetter 1996). To conclude from the failed replication attempts that one cannot trust the results of science would be premature. There are many reasons why the findings of a replication study differ from those of the original study (see Eisend et al. 2016; Lynch et al. 2015):

- Empirical studies usually rely on random sampling, and their results are therefore subject to a *sampling error*. That a replication study based on a particular sample produces a non-significant result may be due to chance. The significant result of the original study can then still be the result that would usually be obtained with several repeated replication attempts. However, based on conflicting results (i.e., based on a significant result from the original study and a non-significant result from the replication study), how can we know whether the original study or the replication study provides the "true" result? For this purpose, the two results can be summarized and integrated by means of a meta-analysis (Lynch et al. 2015; for meta-analysis see Sect. 9.3). If the integrated result is significant, then it confirms the significant result of the original study.
- Some studies cannot be replicated exactly because the *documentation of the methodological details of the original study* is often insufficient to repeat a study in the same detail. One of the reasons for insufficient documentation is that the presentation of studies in many scientific journals is very condensed. Even small deviations from the original study design; for example, the time of day when an experimental study is conducted that measures the emotions or performance of the participants may influence the results of a study.
- In the case of conceptual replications in particular, one attempts to extend the scope of study results beyond the context of the original study by adapting or expanding the original study accordingly. This could be, for instance, a study in a different cultural context, using a different demographic group of people or stimuli other than in the original study. If the results of the replication study differ from the original study, then this may be due to the *contingency of the results.* This means that the results of the original study are valid only in the context of the original study (e.g., in the USA) but not in the context of the (conceptual) replication study (e.g., in Asia).
- Ultimately, of course, there is the possibility that the original study's results were collected and/or analyzed sloppily, or that the researchers even manipulated or falsified the results. In this case, attempts to replicate results may be a way to identify potentially fake results. *Mistakes made by researchers or fabrications of results* tend to limit confidence in science. They also represent a significant ethical problem for science, which Chap. 10 discusses in more detail.

Despite the importance of replication studies to the scientific process, *relatively few replication studies are published.* Evanschitzky et al. (2007) reported a replication rate of 1.2% from 1990 to 2004 in the leading marketing journals (*Journal of Marketing*, *Journal of Marketing Research*, and *Journal of Consumer Research*). That means that only 1.2% of all studies published in these three journals during this period were replication studies. In comparison, replication studies conducted in the period from 1974 to 1989 were at 2.3%, meaning that the replication rate was cut by nearly 50% over time.

Why is it that so few replications are published, even though their importance to science is central? Hunter (2001) highlights two possible reasons for the low interest in replication studies by researchers and journals and mentions the corresponding counterarguments:

- *Low creativity in replication studies*; Counter argument: Sound research requires a solid knowledge base; creativity is not the only criterion for the quality of research.
- *Little increase in knowledge*; Counter argument: A single study with the systematic problems and contingencies of their results is too weak as a knowledge base. Recall the inductive-realistic model of theory testing in Sect. 5.3.

If the chance of getting replication studies published is low, most scientists will have no incentive to conduct the studies. In addition, a replication study that questions a well-published and widely accepted outcome in another study may appear as offensive or even as a personal attack on the authors of the original study. This may also explain why the willingness of researchers to help their colleagues replicate one of their studies is rather low (Reid et al. 1982; Wicherts et al. 2006).

In recent years, marketing research has shown an increased awareness of the need for replication studies. For example, some marketing journals have set up a "Replication Corner" (e.g., *International Journal of Research in Marketing* and *Journal of Marketing Behavior*) or published special issues on replication studies (e.g., *Journal of Advertising*). In special issues of journals or journal sections, replication studies are not in direct competition with original studies, and replication studies thus have a better chance of successfully passing through the peer review process and being published despite the lower creativity of the results. Nevertheless, outside of these outlets for replication studies, the number of replication studies in leading marketing journals has further decreased over the years, while the prevalence of **intra-study replications**, that is, replications of empirical studies within the same project that are actually not considered true replicative research, has increased over years (Kwon et al. 2017).

9.3 Meta-Analysis

A particularly comprehensive and methodologically advanced approach of empirical generalizations is **meta-analysis**. Glass (1976, p. 3) defines a meta-analysis as the "analysis of analyses. . .the statistical analysis of a large collection of analysis results from individual studies for the purpose of integrating the findings." The procedure is to summarize as many as possible (ideally all) relevant empirical results for a particular research question or hypothesis and, to a certain extent, calculate a "common" result, taking into account the different sample sizes. To that end, the presentation of the results from all studies must be comparable and uniform. This occurs by means of so-called **effect sizes**. As explained in Sect. 7.2, effect sizes provide a quantitative assessment of the magnitude of some phenomenon that is used to address a particular research question (Kelley and Preacher 2012). Common effect size measures are correlation coefficients, standardized mean differences, or odds ratios.

Lehmann et al. (1998, p. 746), on the relevance of meta-analyses for empirical marketing research, state:

"One of the most fruitful avenues for analysis is exploring what can be learned from past studies. For example, an advertising agency that has studied the impact of increasing advertising 237 times can learn more from synthesizing the information in the 237 studies than from running the 238th. The process of combining information from past studies is known as empirical generalization and/or meta-analysis (that is the analysis of past analyses). The basic premise is that we can learn from other (past) situations."

The meta-analysis not only integrates results, but also examines their diversity and variability (**heterogeneity**). If the results based on different studies are quite consistent (*homogeneous*), then the overall result that was integrated in the meta-analysis can be regarded as a generalizable finding and reused as such in further research and practice. If the individual results are very different (*heterogeneous*), then this difference can be investigated and (partially) explained in the context of a meta-analysis. This is achieved by applying so-called **moderator** variables (see Sect. 8.2) that are used to explain the variability of effect size values (that is, the realization of a particular effect size measure). Figure 9.1 illustrates this relationship. If there is high variability (heterogeneity) in the effect size values that measure the strength of the relationship between two variables (e.g., attitude toward a product as the independent variable and purchase behavior as the dependent variable), a moderator variable (e.g., product type) might be able to reduce the heterogeneity

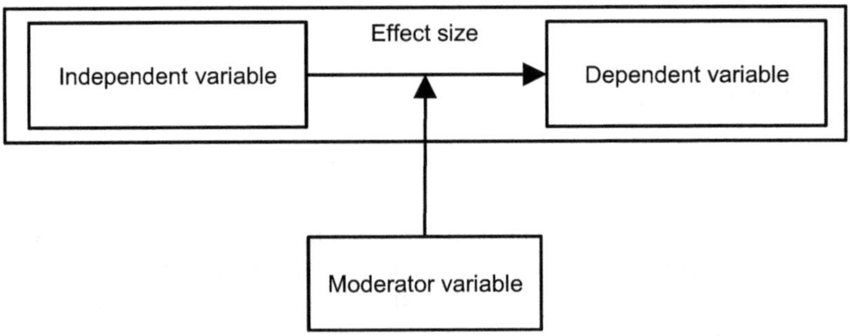

Fig. 9.1 Explaining heterogenous findings in a meta-analysis by means of moderator variables

by explaining under which conditions the effect size values becomes stronger or weaker.

A common moderator variable is the study design, which distinguishes whether the study results were determined in a controlled laboratory experiment or in a field study. If we divide all empirical results into two groups (results from laboratory experiments and results from field studies), we can compare the results from the two groups. If there is a statistically significant difference, we can assume that the overall results cannot be generalized across the different research designs, but they must be differentiated. If there is no statistically significant difference, the study results can be generalized over different study designs. Therefore, the moderator variables might be able to explain the heterogeneity of empirical results. The difference in studies that are integrated in a meta-analysis is therefore not a disadvantage, but rather an advantage because the meta-analysis can show whether the overall result is independent of the specifics of individual studies or if the influence of the differences in the studies is relevant. Moderator variables can refer to different dimensions of generalization. Depending on whether the moderator analysis *reveals homogeneity or heterogeneity of findings* within a particular dimension, the findings can either be *generalized* across that particular dimension (e.g., the findings do not depend on the research method and can be generalized across research methods) or have to be *distinguished* (e.g., the findings depend on the research method and cannot be generalized across research methods). Figure 9.2 illustrates how the assessment of heterogeneity and homogeneity in a meta-analysis can contribute to the generalization of findings across different dimensions.

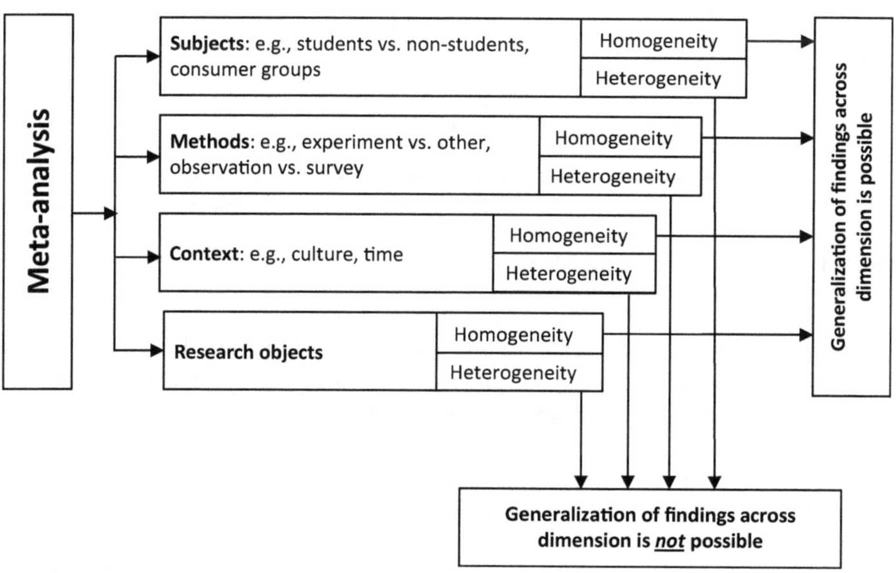

Fig. 9.2 Meta-analysis and generalization across various dimensions

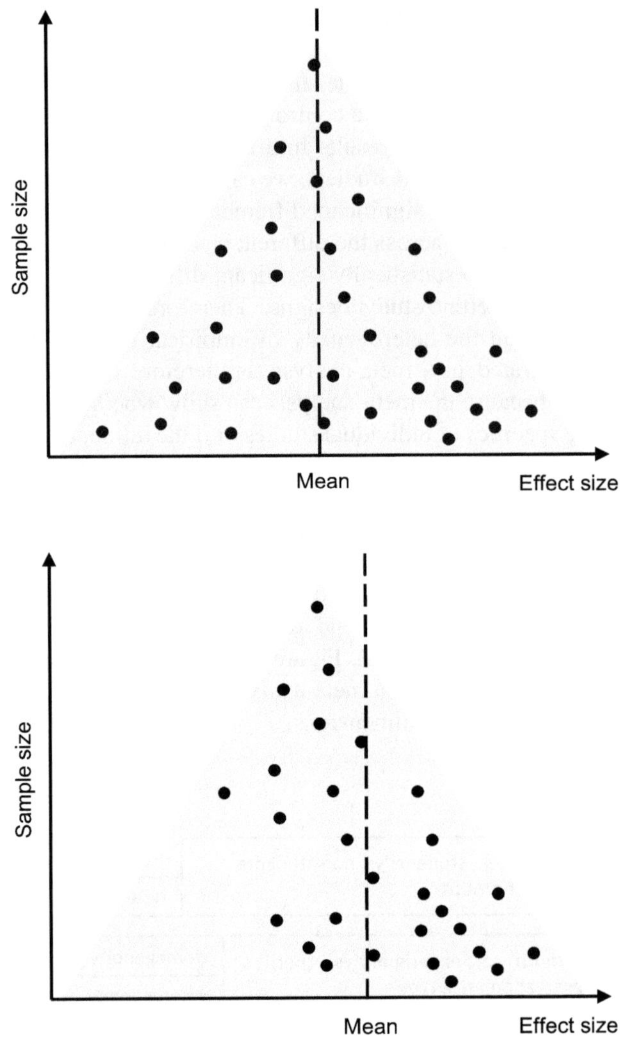

Fig. 9.3 Publication bias and funnel graph

Meta-analyses can help analyze the so-called **publication bias**. This is the frequently empirically confirmed phenomenon that non-significant results are reported less frequently in studies than significant results (for studies on publication bias, see Ferguson and Brannick 2012; Kepes et al. 2012; Renkewitz et al. 2011). Researchers tend to skip reporting insignificant results (see the ethical issue in Sect. 10.2), because these non-significant findings will be less likely to successfully pass the peer review process. As a consequence of publication bias, the results in published studies are upward biased; that is, they are usually "too strong" because the "weak" results are not published at all. In this way, empirical generalizations, which are about the size of an effect, become questionable.

The data in a meta-analysis can be analyzed with regard to the existence of a publication bias and this bias can be corrected. Figure 9.3 illustrates this procedure. In the **funnel graph**, the size of the effect and the sample size of the study from which the effect originates are compared against each other. Very small samples have very large sampling errors, so the variance of the effect size values (e.g., correlation coefficient estimates) around the mean is quite large. The upper funnel graph shows the result of a meta-analysis in which effect size values are scattered, as expected, around the mean according to their sample size. The distribution looks like a triangle or a funnel (hence, the term funnel graph). The lower funnel graph lacks some effect size values, namely, small effect size values based on small samples. These effect size values are more likely to be non-significant compared with large effect size values or effect size values based on large samples. The lower funnel graph thus shows a publication bias: the empirical distribution of the effect size values deviates from the expected distribution in the upper funnel graph. The deviation is systematic, since non-significant findings are missing. The plotted average also illustrates that, in the presence of a publication bias, the integrated (i.e., average) effect found in a meta-analysis is upward biased.

From this funnel graph, it is possible to determine whether there is a publication bias. There is a comprehensive set of methods (documented in detail by Rothstein et al. 2005), with which, among other things, a theoretical distribution can be analytically restored and, thus, an upward biased mean (as shown in the lower graph in Fig. 9.3) can be corrected.

9.4 Generalizability and Theory Development

In many cases, the process of theory development does not take place in such a way that, at the beginning, a "final" theory is formulated and then empirically tested, ultimately leading to rejection or acceptance. Rather, the literature (e.g., Weick 1995) speaks of a **theory continuum** that, in addition to the building of theories and their (empirically-based) acceptance or rejection, also includes phases of modification and refinement of the theories. Essential steps in the theory continuum and the role of generalizations require a brief characterization:

- **Draft**: Theory building is the subject of the fourth chapter and explicitly deals with (in Sect. 4.3.4) the relevance of empirical generalizations. In the *context of discovery* (see Sect. 1.1), generalizations can be used as an inductive approach to theory building. Usually, for the publication of a new theory draft, some preliminary empirical confirmations are needed.
- **Acceptance or rejection** of a theory: These steps are already known from the inductive-realistic model presented in Sect. 5.3. In the case of multiple empirical evidences and confirmations (→ generalization), we decide on an (provisional) acceptance of the theory; if "empirical failures" dominate, we usually reject the

theory. In the *context of justification* (see Sect. 1.1), generalizations help to reduce the risk of errors and mistakes compared with testing a hypothesis in a single study. The point here is that in the case of statements based on a large number of results, it is possible to conclude with greater certainty the "truth" of a theory.

If a theory is broadly accepted after a number of successful tests, and its relevance has been demonstrated, it may be included in relevant *textbooks*. Such textbooks reflect the current status of scientific knowledge concerning a certain field, are the basis for general information and are used in the education of students and young scientists.

Two leading philosophers of science with very different positions characterized the relevance of textbooks for scientific information with the following statements.

Richard Boyd (2002, p. 1) noted:

"For example, if you obtain a good contemporary chemistry textbook you will have good reason to believe (because the scientists whose work the book reports had good scientific evidence for) the (approximate) truth of the claims it contains about the existence and properties of atoms, molecules, sub-atomic particles, energy levels, reaction mechanisms, etc. Moreover, you have good reason to think that such phenomena have the properties attributed to them in the textbook independently of our theoretical conceptions in chemistry."

Thomas Kuhn (1970, p. 43) states:

"Close historical investigation of a given specialty at a given time discloses a set of recurrent and quasi-standard illustrations of various theories in their conceptual, observational, and instrumental applications. These are the community's paradigms, revealed in its textbooks, lectures and laboratory exercises. By studying them and by practicing with them, the members of the corresponding community learn their trade."

- **Refinement**: Section 2.1 characterizes theories by making statements about relationships of concepts with respect to a particular study object (e.g., building customer relationships). Many details of these relationships (e.g., linear or non-linear relationships, magnitude of effects; see Sect. 7.2) are not well-known when the first draft of a theory is designed. This requires numerous detailed studies, the results of which can be summarized and generalized, which then allows statements about typical value ranges of correlations, regression coefficients, etc. In view of the given theoretical framework and with regard to the procedure, such research resembles the "normal science" as characterized and discussed by Thomas Kuhn (1970).
- **Modification**: Modification can be the change of a theory through the addition or elimination of concepts and relationships; this includes moderators and mediators (see Sect. 8.2). For example, if a large number of studies show that a theoretically presumed relationship rarely occurs, this is a reason to rethink the corresponding

variables or the assumed relationships. Moderators influence the strength of the relationship between/among variables; mediators represent the connection between the corresponding variables that are indirectly linked.

In Hunt's inductive-realistic model (2012; see also Sect. 5.3), "acceptance" and "rejection" directly reflect the corresponding "status of the theory"; "refinement" and "modification" are more likely to be assigned to the status of "working acceptance." The latter can also mean that more and clearer results are required for a decision to be made about the acceptance or rejection of the theory.

The process of theory development can be illustrated by the phenomenon of "loss aversion."

Draft: Kahneman and Tversky (1979) introduced the concept of "losses loom larger than gains" and values as assigned by people based on gains and losses, as a critique of the classical economic utility model. They started with simple experiments that showed the enhanced loss sensitivity of consumers.

Refinement: Over the years, many studies have been conducted that attempted to specify, generalize, or modify the concept. For instance, the concept was initially linked to decisions with risks and later extended to riskless choices (Tversky and Kahneman 1991). In marketing, the concept has been used to investigate brand choice of consumers by introducing gains and losses for price variables (Briesch et al. 1997), as well as other quality attributes (Kivetz et al. 2004). Further studies have applied the idea of reference-dependent choices to different areas such as transportation or health care.

Modification: Several variables have been investigated as potential moderators of loss aversion. For instance, to evaluate a product's value, consumers can use either internal reference points (i.e., past information in their memory) or external reference points (i.e., current information provided at the point of purchase). When consumers use external reference points, they show greater price-loss aversion than those using internal reference points (Mazumdar and Papatla 2000).

Acceptance: In 2014, Neumann and Böckenholt presented a meta-analysis and summarized prior research on loss aversion in product choice. They showed how generalizable the concept is and that the degree of loss aversion depends—among other factors—on product and consumer characteristics. In the meantime, the concept of loss aversion is an accepted and important theory that became an integral part of most consumer behavior textbooks (e.g., Hoyer et al. 2018).

How can we describe the role of *generalizations in the research process*? Chapters 4 and 5 focus on the process from theory building to empirical testing; in this chapter, generalizations are more concerned with a larger number of empirical

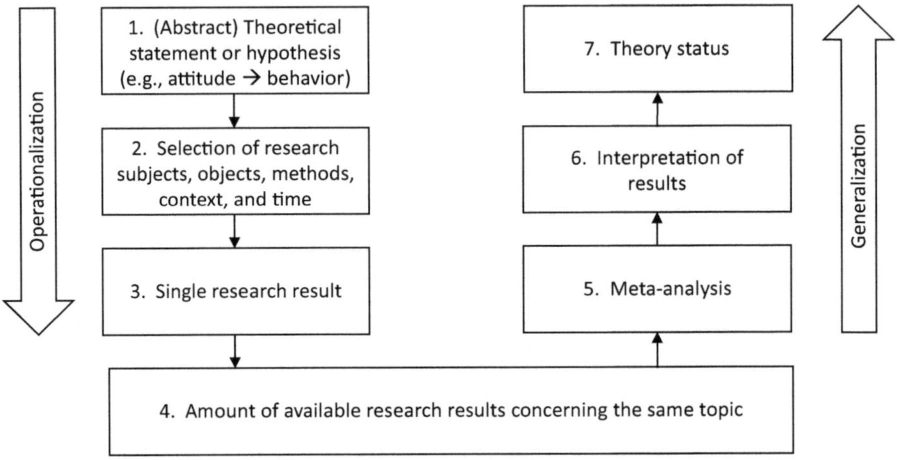

Fig. 9.4 Relationship between operationalization and generalization

findings that help to make more generalized statements (beyond the results of a single study). The following considerations intend to show differences and relationships between the two types of research processes (operationalization and generalization). Figure 9.4 illustrates the relationship.

Operationalization plays a central role in the empirical test of theoretical statements. This is because, in order to verify theories by empirical data, it is necessary to assign measurements to the (abstract) theoretical concepts by using appropriate methods and to analyze the results of these measurements with regard to the suggested hypotheses (see Sect. 6.1). The process of operationalization is, at the same time, a process of concretization and thus of narrowing the research object. For example, in this way, a general question about the relationship between attitude and behavior (Fig. 9.4, step 1) may become a concrete question of the correlation between the attitude toward a particular brand and brand choice. In addition, the corresponding study is performed at a specific time, in a specific context, using certain methods, etc. (Fig. 9.4, step 2), thus leading to a single research result of a specific study dealing with a concrete rather than a general research question (Fig. 9.4, step 3). The question arises as to what meaningfulness such a specific study has for the more general initial question of the generalizability of the research results.

The link to **generalization** is that the result of a single study is no longer considered in isolation, but in relation to other results of studies on the same topic that have already been conducted (meta-analysis) or are generated through replication studies (see Sect. 9.2). This is described by steps 4 and 5 in Fig. 9.4: having a database of the already available study results is the basis for performing a meta-analysis (see previous section). A meta-analysis provides results about the extent to which the theoretically expected relationships have been confirmed (effect sizes) and to what extent changes in variables lead to changes in other variables (magnitude of effects). The interpretation (Fig. 9.4, step 6; see also Sect. 5.2) of such results leads to assessments of the theory status (Fig. 9.4, step 7), that is, the question of whether a theory is accepted, rejected, or in the status of "working acceptance" and therefore should be modified, refined, and tested further. We can see that the process starts with a large number of single results and leads to more general (and more abstract) statements by summarization and integration (using meta-analysis).

Some of the central ideas and their relationships are summarized in Fig. 9.5. At the beginning of the process of theory development is the step of "theory building" (see Chap. 4). Its result is a "theory draft" that is ready for an empirical test (see Chap. 5). A single study (from "hypotheses" to "results") might lead to insights (→ "interpretation"). In many cases, several empirical studies are performed by different researchers and/or in different contexts (e.g. psychology, consumer behavior, behavioral economics). The results of these multiple studies can be collected and analyzed by means of "meta-analysis" (see Sect. 9.3) and interpreted in the next step. Interpretation in this context means to evaluate the existing theory ("theory draft") in the light of the empirical results. The strengthened impact on interpretation provided by *number of results* compared to a single result is symbolized in Fig. 9.5 by a dark bold arrow.

Results of a single study or a meta-analysis reveal empirical successes or failures (see Sect. 5.3; Hunt 2012). If a theory meets a general corroboration due to the dominance of *empirical successes*, some empirical results may be used to modify and/or to refine the theory. Of course, a dominance of *empirical failures* would increase the doubts about the truth of a theory and weakens the theory. Based on this a researcher or the scientific community has to decide whether to reject the theory or not.

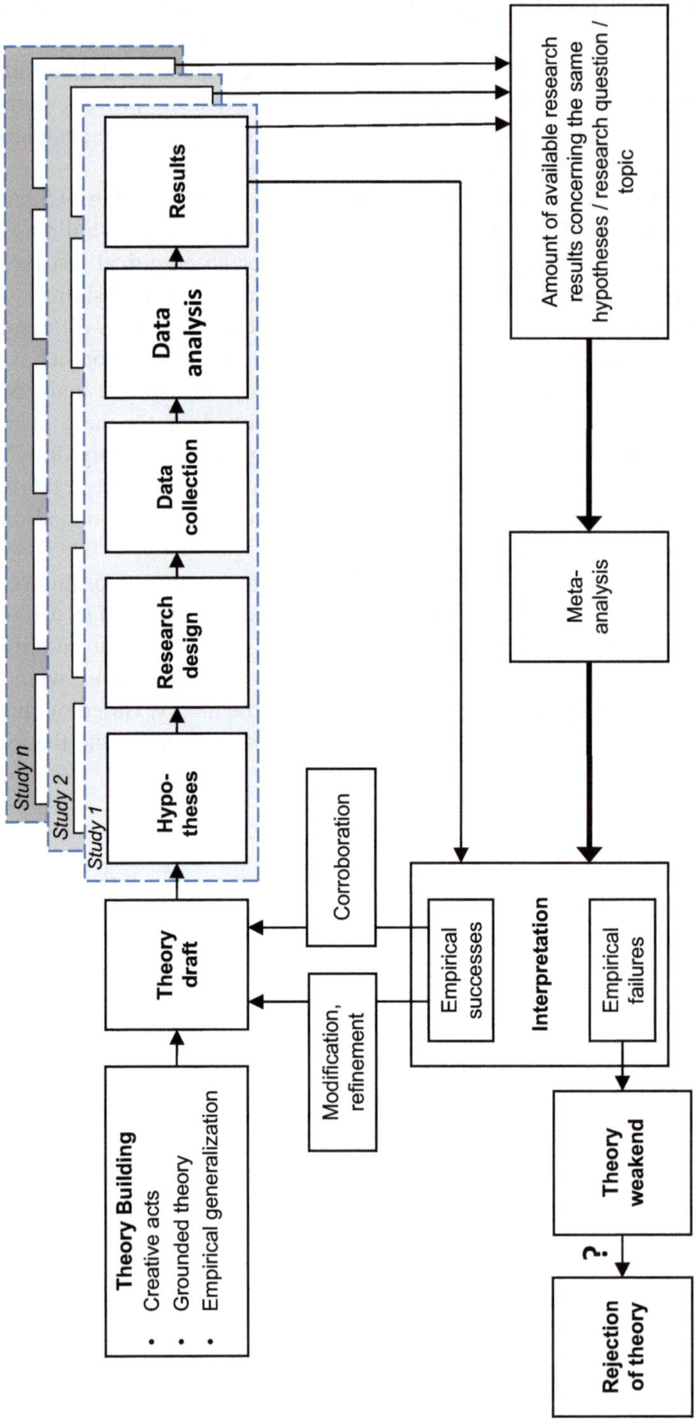

Fig. 9.5 Theory development and empirical research

References

Boyd, R. (2002). Scientific realism. In E. Zalta (Ed.), *The Stanford encyclopedia of philosophy*. http://plato.stanford.edu.

Briesch, R. A., Krishnamurthi, L., Mazumdar, T., & Raj, S. P. (1997). A comparative analysis of reference price models. *Journal of Consumer Research, 24*(2), 202–214.

Cronbach, L. J., Gleser, G. C., Nanda, H., & Rajaratnam, N. (1972). *The dependability of behavioral measurements: Theory of Generalizability for scores and profiles*. New York: Wiley.

Eisend, M., Franke, G. R., & Leigh, J. H. (2016). Reinquiries in advertising research. *Journal of Advertising, 45*(1), 1–3.

Evanschitzky, H., Baumgarth, C., Hubbard, R., & Armstrong, J. S. (2007). Replication research's disturbing trend. *Journal of Business Research, 60*(4), 411–415.

Ferguson, C. J., & Brannick, M. T. (2012). Publication Bias in psychological science: Prevalence, methods for identifying and controlling, and implications for the use of meta-analysis. *Psychological Methods, 17*(1), 120–128.

Glass, G. V. (1976). Primary, secondary, and meta-analysis of research. *Educational Researcher, 5*(10), 3–8.

Hoyer, W. D., MacInnis, D. J., & Pieters, R. (2018). *Consumer behavior* (7th ed.). Boston, MA: Cengage.

Hubbard, R., & Vetter, D. E. (1996). An empirical comparison of published replication research in accounting, finance, management and marketing. *Journal of Business Research, 35*(2), 153–164.

Hunt, S. D. (2012). Explaining empirically successful marketing theories: The inductive realist model, approximate truth, and market orientation. *Academy of Marketing Science Review, 2*(1), 5–18.

Hunter, J. E. (2001). The desperate need for replications. *Journal of Consumer Research, 28*(1), 149–158.

Kahneman, D., & Tversky, A. (1979). Prospect theory: An analysis of decision under risk. *Econometrica, 47*(2), 263–291.

Kelley, K., & Preacher, K. J. (2012). On effect size. *Psychological Methods, 17*(2), 137–152.

Kepes, S., Banks, G. C., McDaniel, M., & Whetzel, D. L. (2012). Publication Bias in the organizational sciences. *Organizational Research Methods, 15*(4), 624–662.

Kerlinger, F., & Lee, H. (2000). *Foundations of behavioral research* (4th ed.). Melbourne: Wadsworth.

Kivetz, R., Netzer, O., & Srinivasan, V. S. (2004). Alternative models for capturing the compromise effect. *Journal of Marketing Research, 41*(3), 237–257.

Kuhn, T. (1970). *The structure of scientific revolutions* (2nd ed.). Chicago: University of Chicago Press.

Kwon, E. S., Shan, Y., Lee, J. S., & Reid, L. N. (2017). Inter-study and intra-study replications in leading marketing journals: A longitudinal analysis. *European Journal of Marketing, 51*(1), 257–278.

Lehmann, D. R., Gupta, S., & Steckel, J. (1998). *Marketing research*. Reading, MA: Addison-Wesley.

Lynch, J. G., Jr., Bradlow, E. T., Huber, J. C., & Lehmann, D. R. (2015). Reflections on the replication corner: In praise of conceptual replications. *International Journal of Research in Marketing, 32*(4), 333–342.

Mazumdar, T., & Papatla, P. (2000). An investigation of reference price segments. *Journal of Marketing Research, 37*(2), 246–258.

McCullough, B. D., & Vinod, H. (2003). Verifying the solution from a nonlinear solver: A case study. *American Economic Review, 94*(1), 873–892.

Neumann, N., & Böckenholt, U. (2014). A meta-analysis of loss aversion in product choice. *Journal of Retailing, 90*(2), 182–197.

Open Science Collaboration. (2015). Estimating the reproducibility of psychological science. *Science, 349*(6251), 943.
Reid, L. N., Rotfeld, H. J., & Wimmer, R. D. (1982). How researchers respond to replications requests. *Journal of Consumer Research, 9*(2), 216–218.
Renkewitz, F., Fuchs, H. M., & Fiedler, S. (2011). Is there evidence of publication Bias in JDM research? *Judgment and Decision making, 6*(8), 870–881.
Rothstein, H. R., Sutton, A. J., & Borenstein, M. (Eds.). (2005). *Publication Bias in meta-analysis. Prevention, assessment and adjustments*. Chichester: Wiley.
Tversky, A., & Kahneman, D. (1991). Loss aversion in riskless choice: A reference-dependent model. *Quarterly Journal of Economics, 106*(4), 1039–1061.
Weick, K. (1995). What theory is not, theorizing is. *Administrative Science Quarterly, 40*(3), 385–390.
Wicherts, J. M., Borsboom, D., Kats, J., & Molenaar, D. (2006). The poor availability of psychological research data for reanalysis. *American Psychologist, 61*(7), 726–728.

Further Reading

Hunter, J. E. (2001). The desperate need for replications. *Journal of Consumer Research, 28*(1), 149–158.
Lipsey, M. W., & Wilson, D. T. (2001). *Practical meta-analysis*. Thousand Oaks, CA: Sage.
Open Science Collaboration. (2015). Estimating the reproducibility of psychological science. *Science, 349*(6251), 943.

Research Ethics and Research Practice 10

10.1 Basic Problems in Research Ethics

Occasionally, reputable newspapers report on "science scandals," which often refer to completely fake research results or plagiarism. These are, of course, very serious cases in which it is made quite clear that such behavior is absolutely unacceptable from an ethical (and oftentimes legal) perspective. In the practice of empirical marketing research, ethical issues often arise at different stages in the research process. These issues can be less serious and less clear; sometimes it is only negligence, but this can have considerable consequences for the scientific process of knowledge generation. Especially in the last 10 years or so, not least because of some prominent cases, the sensitivity for such ethical aspects has grown significantly. For this reason, the first section of this chapter will briefly outline key aspects of research ethics. Section 10.2 characterizes and discusses questions of research ethics that occur during the typical phases of the research process.

First, let's look at some *conflicts* that can lead to ethical questions for researchers. On the one hand, hardly anyone doubts the necessity of ethical principles for scientific research, yet the pressure on scientists has grown so much in recent years (e.g. Honig et al. 2013) that the danger of violating ethical principles has increased:

- For a scientific career, even if it is only a matter of remaining in a scientific profession at all, outstanding publication successes in the leading international journals of the respective discipline are required today.
- In the past, mainly scientists from the US and some European countries published in these few leading journals; the competition for publication opportunities has increased, as more and more authors from around the world try to publish in these journals.
- There is intense competition between journals for reputation and attention (measured primarily by the number of citations of published articles), which

M. Eisend, A. Kuss, *Research Methodology in Marketing*,
https://doi.org/10.1007/978-3-030-10794-9_10

results in publishers and editors being most likely to accept particularly clear and substantial research results for publication.
- In some countries, the pressure factor of "grants" has been added in recent years. In many cases, funders of grants (e.g. interest groups, companies, political institutions, etc.) attach great importance to the fact that the respective projects lead to clear (or seemingly clear, see below) results in a limited amount of time—if possible with the a priori expected results. Otherwise, the chances of successful applications for grants could decrease in the future.

Daniele Fanelli, in several studies, has examined the changes in publication behavior and the possible causes. In one of these studies (Fanelli 2012, p. 891), he found that the proportion of published non-significant results, which are "negative" with regard to the confirmation of a hypothesis, has decreased over time:

"Concerns that the growing competition for funding and citations might distort science are frequently discussed, but have not been verified directly. Of the hypothesized problems, perhaps the most worrying is a worsening of positive-outcome bias. A system that disfavours negative results not only distorts the scientific literature directly, but might also discourage high-risk projects and pressure scientists to fabricate and falsify their data. This study analysed over 4600 papers published in all disciplines between 1990 and 2007, measuring the frequency of papers that, having declared to have 'tested' a hypothesis, reported a positive support for it. The overall frequency of positive supports has grown by over 22% between 1990 and 2007, with significant differences between disciplines and countries."

One of the reasons for preferring "positive" results may be that they are cited more frequently. Another study by Fanelli (2013, p. 701) confirmed this assumption:

"Negative results are commonly assumed to attract fewer readers and citations, which would explain why journals in most disciplines tend to publish too many positive and statistically significant findings. This study verified this assumption by counting the citation frequencies of papers that, having declared to 'test' a hypothesis, reported 'positive' (full or partial) or 'negative' (null or negative) support. Controlling for various confounders, positive results were cited on average 32% more often."

In many cases, empirical research results are not always so smooth and clear as the researchers hoped for:

- Many measurement problems can affect the results.
- In the behavioral sciences, the interaction of a large number of variables is particularly complex, and strong effects of single variables are less common.
- Innovative projects have a higher risk than the progression on known paths.

In this situation, in which the aim is to arrive at clear and original research results, the research process can be very difficult and complex. In some cases (more or less consciously), it may happen that the research process is influenced in order to produce "desirable" results ("**verification bias**"). This is enabled because many details of the research process (e.g. the selection of subjects, the measurements and data preparation) are only limitedly verifiable for outsiders (e.g., reviewers and readers of the publication).

In one of the biggest social science scandals, which centered on the Dutch social psychologist Diederik Stapel, several Tilburg University (Netherlands) committees investigated the numerous data fabrication cases and the methods used and summarized the findings in a comprehensive report (Levelt Committee et al. 2012). This also includes (on p. 48) the following characterization of the so-called *verification bias*:

"One of the most fundamental rules of scientific research is that an investigation must be designed in such a way that facts that might refute the research hypotheses are given at least an equal chance of emerging as do facts that confirm the research hypotheses. Violations of this fundamental rule, such as continuing to repeat an experiment until it works as desired, or excluding unwelcome experimental subjects or results, inevitably tend to confirm the researcher's research hypotheses, and essentially render the hypotheses immune to the facts."

It is important to note that ethics by no means refers only to the extreme cases of fabrication of results or plagiarism (e.g. Martinson et al. 2005). Rather, in the research process, there are many situations—from a research question to a publication—that involve minor or major ethical issues, such as the elimination of certain data ("outliers"), incomplete or selective presentation of results or incorrect information regarding the contribution of several authors in a publication (see Sect. 10.2). Fortunately, the major science scandals uncovering completely fabricated studies or extensive plagiarism rarely occur. Nevertheless, there is evidence of a significantly wider spread of "minor" faults and manipulations in the research process. Table 10.1 shows the results of a survey of more than 2000 psychologists at US universities, who indicated whether they had already used certain questionable approaches in their research practice and to what extent they consider such practices justifiable.

Why have research ethics become so important in science? One might first think of general ethical principles in society, which by all means also apply to scientists and science, namely the rejection of lies, fraud, damage to others and so on. The field of science, however, has some additional specific aspects:

- First of all, *science is free* and not subject to any external control, that is, the correctness of processes and results should be *internally evaluated*, not least by the ethical acceptable behavior of scientists. External control would also be

Table 10.1 Dissemination of questionable research practices (Source: John et al. 2012, p. 525)

Identification of questionable research practices[a]	Proportion of respondents who have already engaged in the respective practice (in %)	Index for the justification of the respective procedure[b]
In a paper, failing to report all of a study's dependent measures	63.4	1.84
Deciding whether to collect more data after looking to see whether the results were significant	55.9	1.79
In a paper, failing to report all of a study's conditions	27.7	1.77
Stopping collecting data earlier than planned because one found the result that one had been looking for	15.6	1.76
In a paper, "rounding off" a *p* value (e.g., reporting that a *p* value of 0.054 is less than 0.05)	22.0	1.68
In a paper, selectively reporting studies that "worked"	45.8	1.66
Deciding whether to exclude data after looking at the impact of doing so on the results	38.2	1.61
In a paper, reporting an unexpected finding as having been predicted from the start	27.0	1.50
In a paper, claiming that results are unaffected by demographic variables (e.g., gender) when one is actually unsure (or knows that they do)	3.0	1.32
Falsifying data	0.6	0.16

[a]Section 10.2 covers the details of problems of such research practices
[b]With a scale with the answer options 0 = "no"; 1 = "possibly"; 2 = "yes" was measured, whether the respective practice is justified. The index represents the mean of these answers

difficult in many areas because of the lack of insight into research processes and specific expertise.

- The central task of science is the *search for truth* and the *avoidance of errors* (Resnik 2008). How can this be ensured if the research process is significantly under the influence of negligence and manipulation?
- It should also be remembered that many fields of research (e.g. life sciences) have *far-reaching consequences* for many people and society at large. Careless work—or even fabricated results—would obviously be completely unacceptable in this regard.
- For science, the exchange of results has central relevance and it would be unthinkable if current research could not be based on past results. In this respect, trust and reliability in science are indispensable.

- Ultimately, it is also about the existence of the scientific system itself, which is largely funded by society (public budgets, foundations, etc.). Sloppy research, fake results and unethical practices would, of course, rightly lead to at least questioning this funding.

In its "Recommendations on Academic Integrity" (Wissenschaftsrat 2015, p. 7), the German Council of Science and Humanities identifies the importance of observing ethical principles for science:

"Honesty, a sense of responsibility and truthfulness are prerequisites in all areas of society and work. Why does science in particular have to make certain of this ethical foundation and continually ensure its stability? Misconduct, fraud and negligence, which can occur in other areas of life, are also possible in science; nonetheless, science has a particular ethical responsibility that compels it to carry out continuous self-monitoring. Science's claim to autonomy—in terms of the freedom of persons and institutions in science—reinforces this ethical responsibility."

Figure 10.1 summarizes key aspects that constitute the area of tension in which scientists are concerned with ethical behavior.

What are the essential ethical principles for scientific research? Resnik (2008, pp. 153ff., 1998, pp. 53ff.) develops some principles that are concretely applicable to

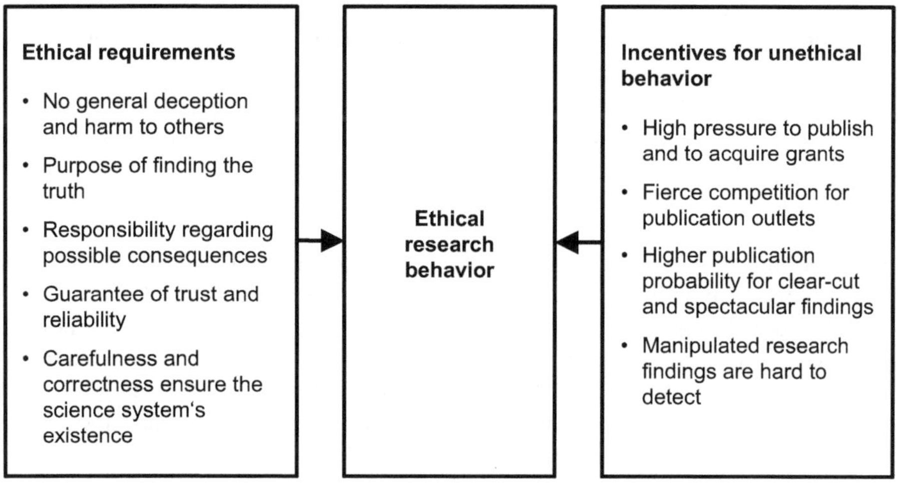

Fig. 10.1 Ethical requirements and incentives for unethical behavior

the respective research practice. Here are the most important of these science-specific principles:

- **Honesty:** "Scientists should practice honesty in research and publication, and in their interactions with peers, research sponsors, oversight agencies, and the public" (Resnik 2008, p. 153).

  Without a doubt, this rather general point concerns almost all ethical requirements for research and it is applicable to the entire research process and the publication of results.
- **Carefulness:** "Scientists should avoid errors in research, especially in presenting results. They should minimize experimental, methodological, and human errors and avoid self-deception, bias, and conflicts of interest." (Resnik 1998, p. 56).

  Carefulness is essential to serve the purpose of research, which is the search for meaningful and true statements. In addition, when using results for further research or for practical applications, it is assumed, of course, that they have been produced with the utmost care.
- **Objectivity**: "Scientists should strive for objectivity in research and publication, and in their interactions with peers, research sponsors, oversight agencies, and the public." (Resnik 2008, p. 153).

  Researchers are sometimes exposed to certain interests (e.g. expectations of success at their home university), which can lead to pressure to obtain certain ("desired") results. However, the goal of objectivity does not only affect the research process in the narrow sense. This should also be applied to reviewers (e.g. in the review process for journals).
- **Openness:** "Scientists should share data, results, ideas, methods, tools, techniques, and resources." (Resnik 2008, p. 153).

  This is about the significant aspect that science can develop only if access to previous knowledge is comprehensively secured. However, openness is often limited in practice in military or commercial research (e.g. market research, pharmaceutical research, etc.). The rising competition in science is another problem.
- **Freedom:** "Scientists should be free to conduct research without political or religious intimidation, coercion, or censorship." (Resnik 2008, p. 154).

  Freedom has been a central "success factor" of scientific research for centuries. Religious- or ideological-influenced research could never have led to the tremendous progress of the past. In Western countries, the freedom of science is largely guaranteed today; however, there are certain limitations, because the allocation of grants and funds can represent the interests of the respective funders.
- **Fair credit allocation:** "Scientists should give credit, but only when credit is due." (Resnik 2008, p. 154).

  Such fairness is the prerequisite for scientific cooperation, not least because the recognition of contributions is of central importance for the professional existence of scientists. Plagiarism as an unmarked takeover of the achievements of other scientists is an extreme example of a violation of this principle. Even the naming of authors who did not have a significant share in the research in question in a publication contradicts the principle. Power relations in the science system can also play a role: "A few decades ago in Germany, it was not uncommon for a professor to publish an article that had been written by an assistant." (Albers 2014, p. 1153).

- **Respect for human subjects:** "Scientists should respect the rights of human subjects and protect them from harm and exploitation." (Resnik 2008, p. 157).

  Adequate behavior toward study subjects has also become a problem in the social sciences, which has led to a number of broadly accepted principles. The standard has become "*informed consent*", which allows the subjects to make a voluntary decision on participation in a study on the basis of appropriate information. In the social sciences, it has also become common for subjects to be protected against damages to their physical or mental health and to be guaranteed about the confidentiality of the data collected (see also Sect. 10.2.3).

In addition, Resnik (2008, p. 154, p. 156) incorporates the following ethical principles, which are less specific to research practice (but not unimportant), in his compilation:

- Respect for colleagues
- Respect for property
- Respect for laws
- Stewardship of research resources
- Social responsibility

Table 10.2 provides some collections of principles on research ethics of several research organizations that are relevant to marketing researchers:

Table 10.2 Statements and principles of research ethics

Organisation	Title	Internet address
Academy of management	"Academy of management code of ethics"	www.aom.org/ethics/
ALL European academies ALLEA	"The European code of conduct for research integrity"	www.allea.org
American Association for Public Opinion Research AAPOR	"AAPOR code of professional ethics and practices"	www.aapor.org
American marketing association	"Statement of ethics"	www.ama.org
American Psychological Association	"Responsible conduct of research"	www.apa.org
European Society for Opinion and Market Research ESOMAR	"ICC/ESOMAR international code on market, opinion and social research and data analytics"	www.esomar.org
Insights association	"Code of standards and ethics for market, opinion, and social research"	www.insightsassociation.org

Ernst-Ludwig Winnacker, former president of the German Science Foundation (2015, p. 23) outlines the consequences of fraud in the science system for fraudsters:

"Of course, there will always be people who are players who take the risk. Then their career is over, and they start a business, living off the money of their parents or their spouse. In any case, a return to the scientific system, where trust is important, will hardly be possible. Anyone who cheats must know that."

The following section illustrates and discusses more concretely ethical problems in the research process. The section follows (roughly) the typical steps of the research process.

10.2 Ethical Issues in the Research Process

10.2.1 Research Topics and Research Questions

The beginning of each empirical study is the determination of the study's topic and the appropriate research questions. Doubts may arise regarding the ethical accountability of certain research objectives. The following example may illustrate this. Consider a market research study designed to develop influencing techniques for children between the ages of 5 and 8 to increase their consumption of (caries-promoting) sweets (e.g. SAGE Editors 2013). Can responsible scientists contribute to "seducing" relatively vulnerable children into harmful behavior? This question is usually answered with "no." But what about more ambiguous cases? Where are the limits?

In 2014–2015, the American Psychological Association (www.apa.org) experienced a fierce controversy over research ethics, as this organization engaged in the development of "enhanced interrogation techniques" (e.g. waterboarding and fake executions) by psychologists commissioned by the American Secret Service or the US military. Hardly anyone will want to ethically justify psychological research into the development of such methods. It is noteworthy that the motives for working with the Ministry of Defense were quite opportunistic, because the military sphere is a major and important employer of psychologists (see Hoffman et al. 2015).

Now, in marketing research, ethical issues are generally not as acute as in some other disciplines. Just think about the very serious discussions on human genetic research, gene modification of agricultural seeds and consequences of nuclear research. Nevertheless, there may also be topics in marketing research in which one should at least ask questions about the ethical justification. Here are some (hypothetical) examples:

- Market entry strategies in international marketing that exclude poor countries from technical or medical progress
- Development of strategies for misleading consumers' price perception
- Impact-maximizing design of misleading advertising
- Questioning the autonomy of consumers through neuromarketing

Often it is not only the avoidance of unethical behavior, but *social responsibility* that is also explicitly required. Resnik (2008, p. 156) formulates this principle in the following way: "Scientists engage in activities that enhance or promote social goods, such as human health, public safety, education, agriculture, transportation, and scientists therefore should strive to avoid harm to individuals and society." This principle can be effective, for example, in scientific opinions on public affairs or warnings about risks from economic developments (e.g. influence of advertising on nutritional behavior). Resnik (2008) cites three arguments that justify the demand for the socially responsible behavior of scientists:

1. Moral obligations that apply in general, including scientists
2. Scientists receive so much support from the public that they should also give something back to society
3. Socially responsible science makes it easier to receive further support from society

Research and teaching at universities is largely funded by public funds. In this respect, it is obvious that not only the perspective of companies can play a role, but also the interests of employees and consumers. Meanwhile, some science organizations have formulated principles of social responsibility for themselves. As an example, see below for the main goals of the University of Bremen. Another example is the Association for Consumer Research, which has a special section titled "Transformative Consumer Research" (TCR): "TCR is a movement within our association that seeks to encourage, support, and publicize research that benefits consumer welfare and quality of life for all beings affected by consumption across the world"(www.acrwebsite.org, accessed July 23, 2018).

Bremen University (Germany) offers an example of positive determination of research goals and the exclusion of certain fields of research (e.g. military research) with its "guiding objectives," from which the following determinations are taken:

"Instructors and students of the University of Bremen are guided by the basic values of democracy, human rights and social justice, which are also the

(continued)

subject of research and teaching in many areas. They will continue to look at the consequences of science in economics, politics and culture and the opportunities for socially and environmentally responsible use of research results (for example, forward-looking technology and economic policy, no military research). The University of Bremen is committed to peace and pursues only civilian purposes." (Source: www.uni-bremen.de/universitaet/profil/leitbild.html, accessed July 23, 2018).

Connections to companies and their associations are obvious and meaningful for marketing research. In some cases, however, there may be attempts to use the special authority of science (see Sect. 1.1) for the interests of individual companies or lobby groups via the allocation of third-party funds, advisory and expert services, company-paid doctoral students and so on, so that they influence the results of scientific research accordingly. The problem of the one-sidedness of paid reports and evaluations is common in scientific, legal and political life. The effort for objectivity of scientific work can be impaired if the preparation of a report for a certain client is associated with considerable payments. If conflicts of interest are possible due to the influence of funders and others, then at least their disclosure in a publication is necessary, which is now a requirement for most scientific journals.

New York University's "Sponsored Research Guidelines" regulate public access to research results that have been funded:

"The University does not conduct or permit its faculty to conduct secret or classified research. This policy arises from concern about the impact of such restrictions on two of the University's essential purposes: to impart knowledge and to enlarge humanity's store of knowledge. Both are clearly inhibited when open publication, free discussion, or access to research are limited. For the same reasons, the University requires that investigators be able to publish the results of their research without prior approval of a sponsor. Agreements may, however, permit sponsors a brief period to review proposed publications and presentations to identify (1) proprietary information that may require patent or copyright protection, or (2) information confidential to the sponsor that must be removed. In general sponsors are granted review periods of 30 to 45 days prior to submission for publication, but review and delay periods should total no more than 90 days". (Source: https://www.nyu.edu/about/policies-guidelines-compliance/policies-and-guidelines/sponsored-research-guidelines.html, accessed July 23, 2018).

The most important criterion for decision-making in such cases is the principle formulated by Schurz (2014, p. 42) that in the case of scientific knowledge, the *context of justification* should be *free* from external influences (see Sect. 1.1). Nevertheless, in the context of discovery and exploitation, in many cases the influences of various interest groups (including the private sector) cannot be completely avoided.

10.2.2 Study Design

The focal point of this phase is the *definition* of a study design and the development of *measurement instruments*. There are usually many options that can significantly influence the results. Of course, this can create a temptation to achieve the most substantial and clear results for favorable publication opportunities (see Sect. 10.1). As an example of the strong influence of the research methodology on results, we can refer to the frequently used survey method in data collection. Numerous studies have shown that even seemingly minor changes in question formulation or questionnaire design can lead to significant differences in results (e.g. Schwarz 1999). The same applies in a similar way to the field of sampling. In the present section, we select and outline some cases that are of widespread importance in research practice.

Ensuring the Validity of Measurements

In the context of this book, the problem of validity of measurements is discussed extensively (see Sect. 6.3). This illustrates the central importance of this aspect. What significance can a study have when it uses data that only insufficiently reflect the theoretically interesting concepts (see Sect. 2.1)? With regard to the lack of validity of erroneous data in the testing of theories, Sect. 3.2 also refers to the discussion of "measurement error underdetermination."

Against this background, a certain amount of evidence for the **validity** of a study is required for its publication in a reputable journal. If validation is a (gradual) exclusion of alternative explanations for the results found (Jacoby 2013, p. 218), then this already suggests that this is a process in which successive tests increasingly provide more certainty of the occurrence of validity. Not all of these tests are reflected in corresponding measures; some are more logical (e.g. in terms of content validity). In addition, there is no established "canon" of validity tests that must be "processed" in each study.

An example of the misuse of validity tests relates to the measurement of "Cronbach's α", which stands for the internal consistency of a multi-item scale, and thus allows statements about the *reliability* (as a necessary condition of validity) of such a scale (see Sect. 6.3). There are some cases in which the process of scale development is such that the items used are extremely similar (or almost identical). Although this contradicts the established principles of scale development, according to which the items should reflect different facets of the measured concept (e.g. Churchill 1979), it favors high α-values and thus increases the chances of publication of a study. Such an approach would be ethically problematic, because the ultimate goal of science, the search for true and meaningful statements (Schurz 2014, p. 19, see also Sect. 1.2) is deliberately disregarded, just to improve the publication chances.

With regard to the validation of measurement instruments, the following principles should at least apply to a responsible research practice:

- Due to the centrality of the validity of study methods, a comprehensive and critical review (and, if appropriate, adaptation) of these methods should be carried out *before* applying these methods in a study (see also Chap. 6).
- The results of a validity check of the methods that are actually used should be *fully documented* in a publication and not limited to a selection of favorable results.

Abusive Use of Pretests
Pretests, above all for checking and improving the measurement methods used (e.g. questionnaires), are today regarded as a standard procedure in empirical research. However, there are also possibilities of abuse insofar as pretests and corresponding changes in the data collection can be made until the desired results come out (again a variant of the "*verification bias*"). Peter (1991, p. 544) comes to a rather skeptical assessment: "It is common practice in some areas of social science to not report such things as how many 'pretests' were done before the desired result was obtained, how many subjects were dropped (...) in order to make the results come out favorably, or how many different manipulations were tried before finding one that worked." Closely related to this is the incorrect practice of including the results of pretests in the publication, depending on whether these results "fit" or not (Laurent 2013).

Lack of Openness in Qualitative Studies
From the perspective of the present book, qualitative studies are most relevant to theory building (see Sect. 4.3.3), and the theories developed become the subject of theory tests (see Chap. 5). In few areas of marketing research, results of qualitative studies are regarded and published as independent research contributions. Qualitative methods are characterized by great openness and freedom in the research process, so that they can support the creative process of theory building (e.g. Creswell 2009; Yin 2011). But if at the beginning of the qualitative research process, the researcher already has more or less defined ideas on the (desired) results, then one must expect that the freedom of the research process would make it relatively easy to achieve these results. If researchers are no longer open-minded about a qualitative research project due to previous theoretical determinations, worldviews or orientation toward the interests of third-party funders, then systematically distorted results, whose causes are hardly recognizable to outsiders, are likely.

10.2.3 Data Collection

This part of the study procedure refers mainly to the process of data collection (e.g. conducting interviews) until the existence of a (still unedited) data set. Central to this is the fair and careful treatment of respondents and test persons. This aspect is of outstanding importance in medical or pharmacological research, but it is by no means a marginal problem for marketing research. In addition, the correct implementation of sampling is important at this stage.

Protection of Respondents or Study Participants

For the participants in empirical studies, who, for instance, complete questionnaires or participate in laboratory experiments, different types of burdens can arise, especially time and stress, and possible disadvantages by disclosing personal information. In the methodology literature (e.g. Shadish et al. 2002, pp. 279ff.; Groves et al. 2009, pp. 375ff.; Rosnow and Rosenthal 2013), there is agreement that the burdens and risks for the study participants must be minimized.

A milestone in the development and implementation of ethical standards for conducting empirical human research was the "*Belmont Report*" (www.hhs.gov), named after the conference venue (Belmont Conference Center, near Baltimore), where in 1978 the National Commission for the Protection of Human Subjects of Biomedical and Behavioral Research set out appropriate policies and guidelines. This was due to experiences from the Nazi era and from the post-war period with unscrupulous experiments on humans, which led to severe damage to the test subjects. Empirical studies in marketing research are usually not associated with such risks, nevertheless, the developed principles also refer to studies in which, at most, *relatively* small disadvantages for the participants may arise. The Belmont Report refers to them as "Basic Ethical Principles":

1. "**Respect for Persons:** Respect for persons incorporates at least two ethical convictions: first, that individual should be treated as autonomous agents, and second, that persons with diminished autonomy are entitled to protection. The principle of respect for persons thus divides into two separate moral requirements: the requirement to acknowledge autonomy and the requirement to protect those with diminished autonomy."
2. "**Beneficence:** Persons are treated in an ethical manner not only by respecting their decisions and protecting them from harm, but also by making efforts to secure their well-being. Such treatment falls under the principle of beneficence. The term "beneficence" often tries to cover acts of kindness or charity that go beyond strict obligation. This document presents beneficence in a stronger sense, as an obligation. Two general rules have been formulated as complementary expressions of beneficent actions in this sense: (1) do not harm and (2) maximize possible benefits and minimize possible harms."
3. "**Justice:** Who ought to receive the benefits of research and bear its burdens? This is a question of justice, in the sense of "fairness in distribution" or "what is deserved." An injustice occurs when some benefit to which a person is entitled is denied without good reason or when some burden is imposed unduly."

It is also important that not all three principals have the same significance for marketing research. Some aspects are more relevant in other contexts, such as medical research (e.g. with regard to new therapies or medicines).

The three "Ethical Principles" have been assigned three more concrete requirements in the Belmont Report:

1. **Informed consent:** The participants agree with the study based on appropriate information on research objectives, possible burdens and data protection. It is therefore up to the requisite "respect for persons" to leave the participants to decide on their participation.
2. **Assessment of risks and benefits:** This aspect corresponds to the principle of "beneficence" because the very benefits of a study (to be maximized) have to be contrasted with the associated (and minimized) burdens. This relation and its possibilities for improvement should be the subject of appropriate considerations in the run-up to the realization.
3. **Selection of subjects:** "The principle of justice gives rise to moral requirements that there be fair procedures and outcomes in the selection of research subjects."

To establish and ensure ethical compliance with human research, *institutional review boards* (IRBs) have been present at US universities and other scientific institutions since 1974 and they must approve the conduct of studies. In many other countries, there are now comparable institutions.

> Here is an example for an Institutional Review Board (IRB) at Northwestern University: (irb.northwestern.edu):
>
> "About the IRB
>
> The protection of research subjects at Northwestern University is a shared responsibility, with the institution, researchers, IRB committees, and the IRB Office working together toward this common goal.
>
> The IRB Office is primarily responsible for developing and directing the University's Human Subject Protection Program (HSPP), which also involves other offices at Northwestern University. The HSPP mission is to be a model program of excellence in protecting the rights and welfare of human subjects involved in research."

Manipulation of Sample Size and Response Rate

Because of the voluntary nature of the participation of respondents or test subjects, a 100% response rate is virtually unattainable in social science studies. Particularly in the academic field (e.g. in studies of doctoral students), there are typically very limited resources. These limited resources aggravate the problem, because often there are no incentives for participation and frequently repeated attempts or reminders are too expensive. For example, a study by Collier and Bienstock (2007) showed that even in studies published in leading international marketing journals, the response rates were usually only less than 50%. It is common that a low level of response rates due to systematic differences between participants and non-participants can lead to biased test results. Here, in the context of research ethics, it is important to critically examine practices that manipulate sample size or response rates to achieve the desired results.

- As we all know, a weak correlation between two variables or a small difference between different groups may be statistically significant if the sample is sufficiently large (see Chap. 7). One tactic for achieving significant results is to increase the sample size accordingly or to combine the data set with other data sets (Levelt Committee et al. 2012; Laurent 2013).
- Even by consciously refraining from higher response rates, one can manipulate results. Laurent (2013, p. 327) formulates a "rule" for such manipulation: "Checking, after the collection of each observation, whether the result is significant (at 5%) and then stopping the data collection immediately, for fear that the result might no longer be significant after additional observations". Of course, he means this as a warning.

Against this background, it is required that the sample size be determined before data collection. "Authors must decide the rule for terminating data collection before data collection begins and report this rule in the article." (Simmons et al. 2011, p. 1362).

10.2.4 Data Preparation and Data Analysis

Typically, after the data collection, a phase of *data preparation* is required, such as identifying wrong records or outliers. Such changes in the data set can be problematic and allow manipulation for desired results. Furthermore, statistical *data analysis* is not as "objective" and independent as it sometimes seems. For example, determining significance levels ($p = 0.01$ or $p = 0.05$ or $p = 0.1$) indeed determines the type and number of "significant" results. An analysis of p-values in leading management journals showed a peculiar accumulation of values just below the usual thresholds of 0.05 and 0.1, respectively, indicating that data have been "processed" just enough to reach those levels of significance (see Albers 2014). An empirical analysis in sociology showed that significantly more p-values were just under than just above the 5% threshold (Gerber and Malhotra 2008), although one would actually expect an approximately even distribution. In the study by Banks et al. (2016), 11% of respondents said they had already manipulated p-values. This is where a relationship with "publication bias," mentioned in Sect. 9.3, becomes apparent: scientific knowledge is systematically distorted if an attempt is made to obtain significant results whenever possible in order to improve the publication chances of a study. In this context, the study by Fanelli (2012), cited in Sect. 10.1, comes to mind.

Ray Fung (2010), from Harvard University, summarizes the results of his research on reported levels of significance in leading management journals:

"Researchers may be dredging through their data to push *p*-values below levels of significance deemed necessary to publish. We examine a random sample of papers from top management articles and compare their hypotheses and *p*-values to a simulated distribution of results that should occur if no data dredging bias exists. Our analysis reveals that data dredging may be occurring. The distribution of *p*-values shows suspicious and statistically significant upswellings preceding the common levels of significance of 0.05 and 0.1. Not a single paper found more than half of its hypothesized results to be nonsignificant, which is statistically infeasible."

Data Manipulation

Without a doubt, the **invention or fabrication of data** is completely unacceptable and usually (on discovery) leads to harsh sanctions, often leading to a loss of the professional position in the science system. Albers (2014) describes corresponding cases. Again, there is a "gray area" of behaviors in which data are not faked, but in which manipulations are made, which may be partly justified and useful, but sometimes also problematic.

On the one hand, the unadulterated reproduction of observations collected in a study is the basis for meaningful empirical results. On the other hand, it may also be useful to eliminate or edit individual records; otherwise, the results would be corrupted. Thus, correlation coefficients or least squares estimates, for example, are influenced in sometimes misleading ways by individual cases with values well beyond the usual range, the so-called "**outliers**" (e.g., Fox 1984, pp. 166–167). The elimination of data also leaves scope for excluding observations from the analysis that "disturb" the desired outcomes (for a full discussion of the problem, see Laurent 2013). After all, in a survey of 344 management researchers by Banks et al. (2016), 29% of respondents said that they had already eliminated cases from data sets to achieve "better" significance values.

Gilles Laurent (2013, p. 326) formulates a general principle for the elimination of outliers:

"In practice, whenever researchers eliminate observations, they should include an appendix that describes precisely their argument for the elimination, as well as the full distribution of observations before and after elimination (. . .)."

"HARKing"

The term "HARKing" ("Hypothezing After the Results are Known", Kerr 1988) is comparable to the term "fishing through a correlation matrix" (Peter 1991, p. 544) and describes a behavior in which the researcher calculates a large number of correlation coefficients, significance tests and so on after the data are available. Then, with such *seemingly* "significant" results, it is possible to come up with "fitting" hypotheses to "enrich" a publication. This does simulate an actually non-existent theoretical basis for these results. In the previously mentioned study by Banks et al. (2016), about 50% of surveyed researchers acknowledged such behavior. The comments in Sect. 7.4 refer to the very limited validity of results obtained in this way.

Nevertheless, it should also be remembered that unforeseen outcomes should not go unacknowledged. There is nothing against an interpretation or discussion of these findings, but the appearance of a theoretically developed and then statistically "successfully" tested hypothesis would be misleading in this case.

Adjustment of Significance Levels and Applied Statistical Methods

Another method to obtain empirical results that (seemingly) confirm the hypotheses that have been theoretically developed is a change in significance levels. Thus, a correlation coefficient or a statistical test at a significance level of $p = 0.05$ may not lead to a significant result but could at $p = 0.1$. One also finds the practice of performing tests with multiple significance levels (e.g. $p = 0.01$, 0.05, and 0.1, respectively). Of course, this increases the proportion of results that are "somehow" statistically significant.

A similar approach is the use of different statistical tests for a particular relationship between variables (Laurent 2013). Since different tests have different properties, the results are usually not identical; and if ethical principles are disregarded, it is often possible—according to the verification bias—to report at least a "suitable" result.

Storage of Data

With regard to the verifiability of test results, today it is increasingly required that the data and documents be kept for a longer period and made accessible as needed. This aspect is important not only in terms of the ability to detect unfair behavior of researchers, but also in terms of performing replication studies and meta-analyses. Chapter 9 outlines their essential importance for the process of generating scientific knowledge.

The storage of data to secure access to it for a certain period of time is not only the task of the authors, but some scientific journals now also take responsibility for this (e.g. in marketing, *Marketing Science* and the *International Journal of Research in Marketing*) and keep this data available to other researchers for replication.

Here is the guideline from the journal *Marketing Science* (pubsonline.informs.org/journal/mksc) regarding the data used in an article that is submitted for publication:

"*Marketing Science* announces its replication policy. Broadly speaking, the policy will require that upon acceptance of a paper by *Marketing Science*, the author(s) of the paper will submit the data and estimation codes used in the paper. The journal will make these files available on its website to scholars interested in replicating accepted paper's results."

10.2.5 Interpretation and Presentation of Results

Between data analysis and publication, the interpretation and presentation of the study results happens, although these steps are certainly overlapping. In research ethics, the focus is on one problem area: the omission of results that do not fit into the overall picture, or the selection of "fitting" results. Thus, the results of the study are incomplete and, in many cases, biased.

Laurent (2013, p. 326) speaks in this context of "*hidden experiments*" or "best of" tactics, meaning the omission of results that do not confirm the central statements of a publication. There is evidence that in some publications, only about half of the original partial studies are reported. In the study by Banks et al. (2016), about 50% of surveyed management researchers stated that they report (or not) on hypothesis testing, depending on significance levels. This may occasionally correspond to the preferences of some reviewers, who are, so to speak, the "gatekeepers" on the way to publication and can exercise corresponding power. Sometimes clear results and short articles are desired, and partial results that do not fit the picture should be left out. However, this is associated with limitations in the search for scientific truth, which Laurent (2013, pp. 326–327) characterized as follows: "If an effect is so weak that it is significant in only four experiments out of eight, this is informative and should be reported. If the effect appears only with certain manipulations, measures, populations, experimental settings, and so forth, this too is informative and should be reported."

Against this background, it is important in a report or publication to fully document the key steps in a study—from the development of measurement instruments and sampling to statistical analysis. This makes it possible for readers and reviewers to comprehend the development of the test results and to critically reflect on them. There are certainly some limits, which the scarcity of space for publications and the patience of readers determine. However, appropriate information can be offered on the Internet or in appendices to larger publications.

Ralph Rosnow and Robert Rosenthal (2013, p. 45) give some advice concerning the transparency, informativeness, precision, accuracy, and groundedness of reporting methods and results:

"By **transparency**, we mean here that the quantitative results are presented in an open, frank, and candid way, that any technical language used is clear and appropriate, and that visual displays do not obfuscate the data but instead are as crystal clear as possible.

By **informativeness**, we mean that there is enough information reported to enable readers to make up their own minds on the basis of the primary results and enough to enable others to re-analyze the summary results for themselves.

The term **precision** is used not in a statistical sense (the likely spread of estimates of a parameter) but rather in a more general sense to mean that quantitative results should be reported to the degree of exactitude required by the given situation.

Accuracy means that a conscientious effort is made to identify and correct mistakes in measurements, calculations, and the reporting of numbers.

Groundedness implies that the method of choice is appropriate to the question of interest, as opposed to using whatever is fashionable or having a computer program repackage the data in a one-size-fits-all conceptual framework."

10.2.6 Publications

Section 10.1 referred to the great, and probably growing, importance of publications in the science system. These are crucial for the opportunity to enter a scientific career and for further development of the career; they significantly influence the chances of success in applying for grants and third-party funding and, in some cases, are the basis for academic honors. The central standards in this respect are the number of publications of a scientist and the quality (degree of innovation, substance, relevance, etc.) of the publications, which often are (simply) assessed on the basis of the status (ranking, reputation, "impact factor" as an indicator for the citation frequency) of the respective journals in which the article is published. Against this background, it is easy to see that scientists are making great efforts and competing to achieve publication success. In a sense, "in the heat of the moment", it can lead to behaviors and practices that are problematic in ethical terms. In the following section, some aspects are addressed from the perspective of the target group of this book (doctoral students, advanced students), all of them potential and future authors. Albers (2014) and Honig et al. (2013) provide further information on the problems of the scientific system and the publication process.

Opportunistic Citation Behavior

Every scientific publication in marketing research is based on an appropriate evaluation of the relevant literature for the respective research subject, in empirical work in particular on the development of the theoretical basis and the presentation and justification of the methodological approach. The bibliography and the corresponding reference list serve to classify the current project and to integrate its results into the development of the field of research, and to adequately acknowledge the achievements of other scholars (see Sect. 10.1), to justify one's own considerations and chosen course of action, and to facilitate access to relevant literature for the readers of the publication. Against this background, the reference list should, of course, focus on sources that are material and somewhat representative of the content of the publication. It appears that there are occasional deviations from this behavior with the aim of increasing publication chances, by citing additional sources that are well appreciated by editors and reviewers of the journal to which the article is submitted. Here are two related practices:

- Adding of citations from publications by members of the "editorial board" of the journal, whose expertise make it likely that they will be considered as reviewers for the submitted article.
- Adding of citations from articles in the journal to which a paper is submitted for publication, but these are not material to the argumentation in the paper. Thus, the author expresses his or her appreciation of this journal and could gain the goodwill of editors and reviewers. Regardless of this, it is not uncommon for an article to be submitted to a thematically highly specialized journal (e.g. *Journal of Product Innovation Management*) that this journal is quoted relatively frequently because of the thematic focus.

In both outlined cases, the authors would mislead the readers to opportunistically increase the publication chances of their article.

Other opportunistic goals are "citation rings" in which scientists within a group (e.g. representatives of a particular research field) cite each other with disproportionate reciprocity, thus increasing one another's fame in the academic world and driving up citation indices. Here also numerous self-citations can play a role. In such cases, other important sources may not be adequately considered, and the information will be withheld from readers.

Plagiarism

In recent years, plagiarism in PhD dissertations by prominent German politicians has attracted a good deal of attention from the general public. Even though this was due to the prominence of the wrongdoers, the fact remains that more or less secretly copying without adequate reference to sources is, according to a very broadly shared view, a completely unacceptable behavior (not only in science). Essentially, it is about the fact that in such cases the use of ideas, results and statements of others in a publication are not adequately identified.

The Academy of Management (2006) gives in the "Code of Ethics" some advice with regard to avoiding plagiarism:

"1. AOM members explicitly identify, credit, and reference the author of any data or material taken verbatim from written work, whether that work is published, unpublished, or electronically available.

2. AOM members explicitly cite others' work and ideas, including their own, even if the work or ideas are not quoted verbatim or paraphrased. This standard applies whether the previous work is published, unpublished, or electronically available."

The sharp rejection of plagiarism in the scientific community is mainly due to the grave violation of the principles of trust, reliability; honesty and fairness (see Sect. 10.1).

"Slicing"

The aforementioned publication pressure on scientists can also lead to attempts to generate as many publications as possible from a larger study. The literature somewhat ironically speaks of "the highest number of publishable units" (Albers 2014, p. 1555) for dividing the results of a project into a larger number of narrowly focused publications. However, the scarce space in the leading journals can also be the reason for the shortest possible publications, in which extensive studies can no longer be given a platform for comprehensive presentation. In such cases, however, all results must be original without repetition of already published results.

What are the ethical problems in this context? First, the question arises as to whether editors as well as reviewers and then the readers of a journal know that several publications have been published or have appeared on various aspects of the project. If not, one gets a distorted impression of the author's contributions in terms of scope and substance. For this reason, it is necessary to state each time an article is submitted whether the results of the respective data have already been published elsewhere. Furthermore, an extensive use of "salami tactics" leads to a waste of scarce space in scientific journals and thus limits the publication possibilities of other studies.

Appropriate Mentioning of the Authors

In view of the already explained relevance of publications for a scientific career, the correct information on authorship is also highly relevant. Appropriate mentioning indicates who is responsible for the published study and has provided the corresponding contribution. The usual rules for naming authors are generally recognized and clear:

- The scientists who have made a *significant contribution* (and only those) should be mentioned as authors. Persons without a contribution should not be named as author. If the contribution of individuals is limited to minor administrative or technical assistance, this can be communicated in a footnote. Sometimes publications refer to co-authors who have not made a direct and significant contribution or worked directly on the project. This can be justified if these individuals have made substantial contributions to enable the project. For example, one could think of scientists who have provided intellectual and administrative contributions to a successful third-party funding application (and thus have designed subprojects) but have not fully cooperated in each subproject. On the other hand, the position of a supervisor at a scientific institution or the supervisor status during a PhD phase does not justify the claim of co-authorship in a publication.

- Normally, the order of authors refers to the proportion of contribution of authors to the publication. If all authors have contributed to approximately the same extent, an alphabetical order (or random order) of names and a corresponding note are common. The hierarchical position does not matter for the order of authors.
- There is no justification for so-called "*ghostwriting*". This is about scientists exploiting the dependencies of others to publish their work under their own name. These are cases of plagiarism (see above), because the "author", who is indicated on the publication, uses a contribution by another person and pretends that it is his or her own achievement.

References

Academy of Management. (2006). *Code of ethics.* aom.org

Albers, S. (2014). Preventing unethical publication behavior of quantitative empirical research by changing editorial policies. *Journal of Business Economics, 84*, 1151–1165.

Banks, G., O'Boyle, E., Pollack, J., White, C., Batchelor, J., Whelpley, C., et al. (2016). Questions about questionable research practices in the field of management: A guest commentary. *Journal of Management, 42*, 5–20.

Churchill, G. A. (1979). A paradigm for developing better measures of marketing constructs. *Journal of Marketing Research, 16*(February), 64–73.

Collier, J., & Bienstock, C. (2007). An analysis of how nonresponse error is assessed in academic marketing research. *Marketing Theory, 7*, 163–183.

Creswell, J. (2009). *Research design—Qualitative, quantitative, and mixed methods approaches* (3rd ed.). Los Angeles: Sage.

Fanelli, D. (2012). Negative results are disappearing from most disciplines and countries. *Scientometrics, 90*, 891–904.

Fanelli, D. (2013). Positive results receive more citations, but only in some disciplines. *Scientometrics, 94*, 701–709.

Fox, J. (1984). *Linear statistical models and related methods*. New York: Wiley.

Fung, R. (2010). *Data anomalies within the management literature.* Working Paper, Harvard University.

Gerber, A., & Malhotra, N. (2008). Publication Bias in empirical sociological research—Do arbitrary significance levels distort published results? *Sociological Methods and Research, 37*, 3–30.

Groves, R., Fowler, F., Couper, M., Lepkowski, J., Singer, E., & Tourangeau, R. (2009). *Survey methodology* (2nd ed.). Hoboken, NJ: Wiley.

Hoffman, D. H., Carter, D. J., Viglucci Lopez, C. R., Benzmiller, H. L., Guo, A. X., Yasir Latifi, S., et al. (2015). *Report to the special Committee of the Board of directors of the American Psychological Association: Independent review relating to APA ethics guidelines. National Security Interrogations, and torture*. Chicago, IL: Sidley Austin. Accessed September 30, 2018, from http://www.apa.org/independent-review/revised-report.pdf

Honig, B., Lampel, J., Siegel, D., & Drnevich, P. (2013). Ethics in the production and dissemination of management research: Institutional failure or individual fallibility? *Journal of Management Studies, 51*, 118–142.

Jacoby, J. (2013). *Trademark surveys – Designing, implementing, and evaluating surveys*. Chicago: American Bar Association.

John, L., Loewenstein, G., & Prelec, D. (2012). Measuring the prevalence of questionable research practices with incentives for truth telling. *Psychological Science, 23*, 524–532.

Kerr, N. (1988). HARKing: Hypothesizing after the results are known. *Personality and Social Psychology Review, 2*, 196–217.

Laurent, G. (2013). Respect the data! *International Journal of Research in Marketing, 30*, 323–334.

Levelt Committee, Noort Committee, & Drenh Committee. (2012). *Flawed science: The fraudulent research practices of social psychologist Diederik Stapel.* Tilburg University. Accessed July 3, 2018, from www.tilburguniversity.edu

Martinson, B., Anderson, M., & de Vries, R. (2005). Scientists behaving badly. *Nature, 435*, 737–738.

Peter, J. (1991). Philosophical tensions in consumer inquiry. In T. Robertson & H. Kassarjian (Eds.), *Handbook of consumer behavior* (pp. 533–547). Englewood Cliffs, NJ: Prentice-Hall.

Resnik, D. (1998). *The ethics of science*. London: Routledge.

Resnik, D. (2008). Ethics of science. In S. Psillos & M. Curd (Eds.), *The Routledge companion to philosophy of science* (pp. 149–158). London: Routledge.

Rosnow, R. L., & Rosenthal, R. (2013). Quantitative methods and ethics. In T. D. Little (Ed.), *The Oxford handbook of quantitative methods in psychology* (pp. 32–54). Oxford/New York: Oxford University Press.

SAGE Editors (Ed.). (2013). *SAGE brief guide to marketing ethics*. Los Angeles: Sage.

Schurz, G. (2014). *Philosophy of science. A unified approach*. London: Routledge.

Schwarz, N. (1999). Self-reports—How questions shape the answers. *American Psychologist, 54*, 93–105.

Shadish, W. R., Cook, T. D., & Campbell, D. T. (2002). *Experimental and quasi-experimental designs for generalized causal inference*. Boston, MA: Houghton Mifflin.

Simmons, J., Nelson, J., & Simonsohn, U. (2011). False-positive psychology: Undisclosed flexibility in data collection and analysis allows presenting anything as significant. *Psychological Science, 22*(11), 1359–1366.

Wissenschaftsrat. (2015). *Empfehlungen zu wissenschaftlicher Integrität*. Accessed July 3, 2018, from www.wissenschaftsrat.de

Yin, R. (2011). *Qualitative research from start to finish*. London: Guilford.

Further Readings

Laurent, G. (2013). Respect the data! *International Journal of Research in Marketing, 30*, 323–334.

Resnik, D. (1998). *The ethics of science*. London: Routledge.

Rosnow, R., & Rosenthal, R. (2013). Quantitative methods and ethics. In T. Little (Ed.), *The Oxford handbook of quantitative methods* (Vol. 1, pp. 32–54). Oxford/New York: Oxford University Press.

Shadish, W. R., Cook, T. D., & Campbell, D. T. (2002). *Experimental and quasi-experimental designs for generalized causal inference*. Boston, MA: Houghton Mifflin.

Index

M. Eisend, A. Kuss, *Research Methodology in Marketing*,
https://doi.org/10.1007/978-3-030-10794-9

MIX
Papier aus verantwortungsvollen Quellen
Paper from responsible sources
FSC® C105338

If you have any concerns about our products,
you can contact us on
ProductSafety@springernature.com

In case Publisher is established outside the EU,
the EU authorized representative is:
Springer Nature Customer Service Center GmbH
Europaplatz 3, 69115 Heidelberg, Germany

Printed by Libri Plureos GmbH
in Hamburg, Germany